HUMAN-BUILT WORLD

science.culture

A series edited by Steven Shapin

Other science.culture series titles available:

HUMAN-BUILT WORLD

How to Think about Technology and Culture

Thomas P. Hughes

THE UNIVERSITY OF CHICAGO PRESS
CHICAGO AND LONDON

Thomas P. Hughes is the Mellon Professor Emeritus in the
Department of the History and Sociology of Science at the University
of Pennsylvania and Distinguished Visiting Professor, Massachusetts Institute
of Technology. He is the editor or coeditor of seven books and author of
four books, including *American Genesis: A Century of Invention and
Technological Enthusiasm, 1870–1970*, which was a finalist
for the Pulitzer Prize.

The University of Chicago Press, Chicago 60637
The University of Chicago Press, Ltd., London
© 2004 by The University of Chicago Press
All rights reserved. Published 2004
Printed in the United States of America

13 12 11 10 09 08 07 06 05 04   1 2 3 4 5
ISBN: 0-226-35933-6 (cloth)

Library of Congress Cataloging-in-Publication Data

Hughes, Thomas Parke.
    Human-built world : how to think about technology and culture /
Thomas P. Hughes.
        p. cm. — (Science.culture)
Includes bibliographical references and index.
    ISBN 0-226-35933-6
    1. Technology—Social aspects—United States.   2. Technology—
United States.   I. Title.   II. Series.
T14.5.H84 2004
303.48'3—dc22

2003018426

♾ The paper used in this publication meets the minimum requirements
of the American National Standard for Information Sciences—
Permanence of Paper for Printed Library Materials,
ANSI Z39.48-1992.

Contents

Illustrations

Acknowledgments

First presented as the Page Barbour Lectures at the University of Virginia in 1995, I have extensively revised the original to transform it into the *Human-Built World*. Presentations of chapters to audiences at the Cosmos Club in Washington, D.C., the Dudley House at Harvard University, Helsinki University, University of Utah (Gould Lecture), MIT (Morison Lecture), Norwegian School of Management, University of Manchester (Caldwell Lecture), and Church of Saint Martins in the Fields (the Forum) have stimulated various revisions.

My late wife, Agatha Chipley Hughes, counseled me and edited the original version of the manuscript. Subsequently, my daughter, Agatha Heritage Hughes, and my son, Lucian Hughes, to whom this book is dedicated, provided invaluable editorial suggestions and support. The endorsement of Stephen Shapin, the distinguished editor of the series in which *Human-Built World* appears, has encouraged me greatly. For enthusiastic and knowledgeable editorial support and advice, I am indebted to executive editor Susan Abrams, editor Christie Henry, editorial associate Jennifer Howard, and senior manuscript editor Erin DeWitt of the University of Chicago Press, a publisher of impressively high standards and integrity. For invaluable assistance in gathering illustrations and permissions, I thank Mary Kathryn Hassett and Betsy Brickhouse, and for the index, Jan Williams. Several unnamed reviewers of the manuscript provided guidance for revisions.

I am grateful for discussions with, comments of, and encouragement from my professional colleagues and friends Fred Allen, Mary Anderson, Nancy and Steven Bauer, Barbara Butler, Bernard Carlson, Edward Constant II, Paul Edwards, Ann Greene, Gabrielle Hecht, David Hounshell, Arne Kaijser, Evelyn Fox Keller, Philip Khoury, Timothy Lenoir, Svante Lindqvist, Carolyn Marvin, Leo Marx, Victor McElheny, Everett Mendelsohn, David Mindell, Stanislaus von Moos, Joel Moses, Joachim Nettelbeck, David Nye, Glenn Porter, Fred Quivik, Denise Scott Brown, Gino and Bettina Segré, Anne Spirn, John Staudenmaier, Jane Summerton, Robert Tate, Lars Thue, Frank Trommler, Robert Venturi, David Warsh, Rosalind Williams, and Thomas Zeller. Nancy Essig of the University of Virginia Press gave support and advice during the early phase of this book project. Neal A. Hébert, Sybil Csigi, Patricia Johnson, Debbie Meinbresse, and Joyce Roselle provided administrative support.

I am also indebted to a number of libraries and archives, especially the Hagley Museum and Library, Wilmington, Delaware; the Van Pelt Library at the University of Pennsylvania; the Alderman Library at the University of Virginia; and the Massachusetts Institute of Technology Archives and Library.

The Andrew W. Mellon Foundation, especially its senior vice president Harriet Zuckerman, has generously funded my research and writing over a decade. I am also indebted to the University of Pennsylvania, whose liberal retirement policies help make it possible for me to continue my scholarly activities in retirement.

Finally, I am deeply appreciative of the encouragement and wise counsel of my good and longtime friend Mary Hill Caperton.

CHAPTER ONE

Introduction: Complex Technology

Technology is messy and complex. It is difficult to define and to understand. In its variety, it is full of contradictions, laden with human folly, saved by occasional benign deeds, and rich with unintended consequences. Yet today most people in the industrialized world reduce technology's complexity, ignore its contradictions, and see it as little more than gadgets and as a handmaiden of commercial capitalism and the military. Too often, technology is narrowly equated with computers and the Internet, which are mistakenly assumed to have been invented and developed in a private-enterprise market context. Having cultivated technology impressively, Americans, especially, need to understand its complex and varied character in order to use it more effectively as means to a wide variety of ends. Both the Flying Fortresses of World War II and the flying buttresses of the Middle Ages are technological artifacts.

In the following chapters, I draw upon and summarize the ideas of public intellectuals, historians, social scientists, engineers, natural scientists, artists, and architects who have helped me over five decades to better understand the complexity of technology and its multiple uses. Because the series in which this book appears is about science, technology, and culture, I also include descriptions of the works of artists and architects who have influenced my view of technology.

Since most of the works considered were done decades, even centuries, ago, they provide a historical perspective. This helps me and I hope my readers to move out of our present mind-set and consider the different ways in which technology has been envisioned as a means to solve problems, some different, some resembling those we face today. History does not repeat itself in detail, but drawing analogies between past and present allows us to see similarities. For this reason, generals study military history, diplomats the history of foreign affairs, and politicians recall past campaigns. As creatures in a human-built world, we should better understand its evolution.

Defining Technology

Defining technology in its complexity is as difficult as grasping the essence of politics. Few experienced politicians and political scientists attempt to define politics. Few experienced practitioners, historians, and social scientists try to inclusively define technology. Usually, technology and politics are defined by countless examples taken from the present and past. In the case of technology, it is usually presented in a context of usage, such as communications, transportation, energy, or production.

The word "technology" came into common use during the twentieth century, especially after World War II. Before then, the "practical arts," "applied science," and "engineering" were commonly used to designate what today is usually called technology. The *Oxford English Dictionary* finds the word "technology" being used as early as the seventeenth century, but then mostly to designate a discourse or treatise on the industrial or practical arts. In the nineteenth century, it designated the practical arts collectively.

In 1831 Jacob Bigelow, a Harvard professor, used the word in the title of his book *Elements of Technology . . . on the Application of the Sciences to the Useful Arts.* He remarked that the word could be found in some older dictionaries and was beginning to be used by practical men. He used "technology" and the "practical arts" almost interchangeably, but distinguished them by associating technology with the application of science to the practical, or useful, arts. For him, technology involved not only artifacts, but also the processes that bring them into being. These processes involve invention and human ingenuity. In contrast, for Bigelow, the sciences consisted of discovered principles, ones that exist independently of humans. The sciences are discovered, not invented.

I also see technology as a creative process involving human ingenuity. Emphasis upon making, creativity, and ingenuity can be traced back to *teks*, an Indo-European root of the word "technology." *Teks* meant to fabricate or to weave. In the Greek, *tektön* referred to a carpenter or builder and *tekhnë* to an art, craft, or skill. All of these early meanings suggest a process of making, even of creation. In the Middle Ages, the mechanical arts of weaving, weapon making, navigation, agriculture, and hunting involved building, fabrication, and other productive activities, not simply artifacts.

Landscape architect Anne Whiston Spirn's definition of landscape in *The Granite Garden: Urban Nature and Human Design* (1984) suggests a way of thinking about technology. For her, landscape connects people and a place, and it involves the shaping of the land by people and people by the land. The land is not simply scenery; it is both the natural, or the given, and the human-built. It includes buildings as well as trees, rocks, mountains, lakes, and seas. I see technology as a means to shape the landscape.

As noted, "technology" was infrequently used until the late twentieth century. When a group of about twenty American historians and social scientists formed the Society for the History of Technology in 1958, they debated whether the society should be known by the familiar word "engineering" or the unfamiliar one "technology." They decided upon the latter, believing "technology," though the less used and less well-defined term, to be a more inclusive term than "engineering," an activity that it subsumes.

So historians of technology today are applying the word to activities and things in the past not then known as technology, but that are similar to activities and things in the present that are called technology. For example, machines in the nineteenth century and mills in the medieval period are called technology today, but they were not so designated by contemporaries, who called them simply machines and mills.

In 1959 the Society for the History of Technology began publication of a quarterly journal entitled *Technology and Culture*. The bewildering variety of things and systems referred to as technology in the journal's first two decades reveals technology's complex character. Rockets, steam and internal combustion engines, machine tools, textiles, computers, telegraphs, telephones, paper, telemetry, photography, radio, metals, weapons, chemicals, land transport, production systems, agricultural machines, water transport, tools, and instruments all appear as technology in the journal's pages. Yet the various kinds of technology noted in *Technology and Culture* have a common denominator—most can be associated with the creative activities, individual and collective, of craftsmen, mechanics, inventors, engineers, designers, and scientists. By limiting technology to their creative activities, I can avoid an unbounded definition that would include, say, the technology of cooking and coaching, as widespread as they may be.

Having taught the history of technology for decades and having faced the difficulties of defining it in detail, I have resorted to an overarching definition, one that covers how I use the term in the following chapters. I see technology as craftsmen, mechanics, inventors, engineers, designers, and scientists using tools, machines, and knowledge to create and control a human-built world consisting of artifacts and systems associated mostly with the traditional fields of civil, mechanical, electrical, mining, materials, and chemical engineering. In the twentieth and twenty-first centuries, however, the artifacts and systems also become associated with newer fields of engineering, such as aeronautical, industrial, computer, and environmental engineering, as well as bioengineering.

Besides seeing technology associated with engineering, I also consider it being used as a tool and as a source of symbols by many architects and artists. This view of technology allows me to stress the aesthetic dimensions of technology, which unfortunately have been neglected in the training of engineers, scientists, and others engaged with technology.

My background helps explain why I have chosen a definition emphasizing creativity and control. Before earning a Ph.D. in modern European history, I received a degree in mechanical and electrical engineering. In the 1950s, I found engineering and related technology at their best to be creative endeavors. Not uncritical of their social effects, I still considered them potentially a positive force and expressed a tempered enthusiasm for them and their practitioners.

Since then, I have learned about the Janus face of technology from counterculture critics, environmentalists, and environmental historians. Yet the traces of my enthusiasm still come through in my publications, especially this one. Hence my defining technology as a creative activity, hence my willingness to sympathetically portray those who have seen technology as evidence of a divine spark, and hence my interest in those who consider the machine a means to make a better world. Yet this sympathetic view is qualified by what I have learned from critics of technology.

Overarching Theme: Creativity

Despite the varied approaches to technology taken by the authors, artists, and architects considered in this book, I find several overarching and related themes emerging as my account moves from the past toward the present. Most of my sources see technology, as I do, offering creative means to a variety of ends. They explore the various uses of technology. They also understand it to be especially important as means to create and to control a human-built world, the extent of which is steadily increasing. As a result, my sources have given me and should give my readers a far greater appreciation of our responsibilities for the use of technology and for the characteristics of the human-built world it creates.

Creativity is usually associated with the fine arts and architecture, yet technology throughout history has enabled humans to exercise godlike creative powers. For the Greeks, Prometheus symbolized creativity in stealing fire from the gods, not with paint and canvas. Conventionally seen as a gifted artist, Leonardo da Vinci presented himself to the world as an architect-engineer who filled his notebooks with inventions and engineering projects, including canals, automated textile machines, and machine tools. Johann Wolfgang von Goethe in his epic poem *Faust* had the protagonist ultimately finding earthly fulfillment in the creation, by the drainage of wetlands, of new land upon which humans will thrive.

Today in a secular age dedicated to a consumer culture, we do not see technology in the grand perspective suggested by Leonardo and Goethe. We are content to let inventors and entrepreneurs, energized by market forces, lay claim to the laurels of creativity. During the past century, American inventors became heroic figures symbolizing the country's aptitude for innovation. Thomas Edison remains the foremost independent inventor-entrepreneur remembered for his electric light, phonograph, telegraph, and numerous other inventions. He was the proverbial small-town, plainspoken, self-made American whose untutored genius brought him fame and wealth. How unlike the highly complex Leonardo, who was both artist and engineer.

We should not be surprised that a century later Americans eager to recapture and refresh their image as a nation of inventors relish stories about a new wave of youthful inventor-entrepreneurs preparing business plans to finance Silicon Valley start-up companies that are intended to stimulate a computer revolution. Goethe's Faust would hardly have asked the moment of creation to linger, if the result was simply one more consumer good.

Along with inventors, engineers are also seen today as the creators of the human-built world. During the last century, Americans and Germans pictured engineers as robust, commonsensical, practical, self-made, rather dull men. With the late-nineteenth-century rise of engineering schools, an engineering education became the means for young men of humble origins to move up into the ranks of industrial managers and preside over a market-driven industrial scene. Only within the past few decades have women in substantial numbers been encouraged to enroll in engineering schools and to enter the profession. Perhaps they will introduce more complexity into a profession inclined toward reductionism.

Overarching Theme: Human-Built World

According to the myth of creation found in the book of Genesis, humans have been engaged in creating a living and working place, a human-built world, ever since their ouster from the Garden of Eden. There, Eve and Adam did not have to provide food, clothing, or shelter for themselves. Subsequently humans used technology to transform an uncultivated physical environment into a cultivated and human-built one with all of its artifacts and systems.

The transformation has been especially rapid and obvious during industrial revolutions beginning in England in the eighteenth century, extending to the United States and Germany in the nineteenth, and into Japan and other regions of the developing world in the twentieth. In order to define the concept of the human-built, I could focus upon history in countless times and places, but a brief survey of the American experience should suffice. By 1900 even Europeans acknowledged that the United States had become technology's nation.

In the traditional account, Americans transformed an uncultivated wilderness into a living and work place—the human-built world. In recent years, however, historians have revealed that at least a million Native Americans inhabited the so-called wilderness of the seventeenth century and they had changed the landscape by controlled burning of grasslands, clearing of forests, and the planting of crops.

There are no better defining examples of the human-built than the industrial metropolises that mushroomed in the United States during the late nineteenth century. Young men and women who knew exhausting labor of the farms flocked into expanding industrial cities like Boston, New York, Philadelphia, Baltimore, Cleveland, and Chicago with their diversions and promise of blue- and white-collar work.

Living in a human-built urban world shaped American character, as my sources will testify. Unlike Frederick Jackson Turner, an eminent historian, who in his essay "The Significance of the Frontier in American History" (1893) wrote persuasively of the influence of the natural frontier on American character, I am more interested in the ways in which the human-built environment has shaped character. Early in the nineteenth century, Turner asked his readers to stand at Kentucky's Cumberland Gap and watch successive waves of Indians, fur traders, hunters, cattle raisers, and pioneer farmers flowing westward. He believed that frontier experiences brought them and later Americans to favor political democracy, cross-pollination of ethnic groups, inquisitiveness, materialism, and individualism.

Instead of natural frontier images, I seek, through my sources, images capturing the essence of the human-built world. It can be observed on the streets of industrial Chicago in 1910 populated by skilled blue-collar workmen, eager young inventors, aspiring white-collared managers, and upwardly mobile, clean-shaven farm boys become mechanical engineers. There were also young women born and bred in small towns finding employment as clerks in the newly established department stores, as stenographers in places of business, and as laborers in manufacturing shops. The tumultuous pace and the callous demands of a commercial culture overwhelmed some young men and women who drifted into the urban slums or into prostitution or petty crime.

Many of them flourished, however, as their country—once despised by industrial Britain as rural and uncouth—became the world's preeminent industrial and technological power. A construction site for centuries, America spawned a nation of builders. The business of America was building. The spirit of the people was not only free enterprise, but also demiurge—the spirit of the Greek god who created the material world.

1. Not a tree or a blade of grass in human-built Chicago, ca. 1910. Courtesy of the Chicago Historical Society.

Seeing a human-built world around them, engineers, scientists, and managers believed that they had the creative technological power to make a world according to their own blueprints. They considered the natural world as expendable and exploitable or simply as scenery. Railroads, highways, telegraphs, and telephones allowed transportation and communication systems to reach the remote hinterland. Electric power energized factories far from the rail networks serving industry dependent upon coal. Artificial fertilizers and agricultural machinery increased the yield of lands once exhausted. Chemists transformed matter and provided abundance where nature had denied resources, even though the side effects were hidden for decades. The creative spirit, the desire to create a human-built world, was reaching its apogee.

My sources develop not only upon the overarching themes of

creativity and the human-built, but also subthemes dealing with the varied ends that technology can serve. I develop these subthemes in individual chapters that are arranged chronologically. Before discussing the subtheme explored in a chapter, I provide context by surveying technological developments taking place during the periods in which the subtheme most obviously emerged.

Second Creation

Chapter 2, for instance, focuses upon the nineteenth-century transformation of the America landscape into a human-built world by European settlers and African slaves. The chapter examines the expressed motivations, many of European origin, of the settlers. Economic and political forces drove them, but many also saw themselves motivated by a divine mission.

Seeing technology within a religious context and as serving religious ends, the American settlers provide a starkly differing perspective on technology than the one prevailing today in the industrialized world. For them, the transformation of a howling wilderness into a living and work place was more than a utilitarian venture. They believed that technology, a gift of God, could help them transform the New World into a Garden of Eden. They called this marvelous transformation a "second creation."

Living in times when religion provided for many a defining perspective on the world, settlers believed that the creative act depended upon a divine spark that God, the author of the first creation, had placed in his creatures. Many early Americans eagerly accepted the injunction in the book of Genesis to subdue the earth and have dominion over all creatures. Believing themselves so empowered, they often fell into the habit of exploiting nature for their immediate material gain and failed to accept other biblical injunctions demanding that they act as nurturing stewards of God's and their own creations.

On the other hand, some influential Americans cared deeply

about the moral and environmental character of the second creation. They wished to retain and maintain a natural environment
that they found sustaining, an environment that the aggressively ex-
ploitative called a wilderness. Other contemplative Americans nurtured a vision of a middle landscape, a garden, that mediated between unimproved nature and the human-built. Still others of a
philosophical cast of mind asked that positive benign values be embedded in the human-built world.

Machine for Production

I show in chapter 3 that the effort to bring about a second creation
did not culminate in a new Eden, but, instead, in an industrial revolution that brought a human-built world consisting of steel, electric power, internal combustion engines, chemically transformed
materials, and a plethora of consumer goods. By the early twentieth
century, American technology had become the most prolific machine for production that the world had ever known. It partially
fulfilled the material needs of millions of Americans escaping from
poverty, but it also imposed demanding working and living conditions upon them.

American commentators reacted positively to this industrial revolution and its outpouring of consumer goods. Their reactions became embedded in American culture and their values continue to
sustain consumerism today, even among those satiated with goods.
Some American spokespersons became so infatuated with machines
that they spoke metaphorically and approvingly of the machinery
of government and machinery of the body. Believing machines not
only to be fruitful and efficient, contemporaries also took them to be
easily controllable.

Because a similar revolution occurred in Germany about the
same time, I have taken this as an opportunity to consider Germans
whose reactions to technology as a machine for production have
substantially influenced my concept of technology. German intel-

lectuals reacted to mechanization more critically than did Americans, however. For instance, Oswald Spengler, a popular early-twentieth-century German historian, had ideas about technology that are virtually unthinkable in twenty-first-century America. He blamed technology for the decline of the West and for stimulating among Westerners the sin of human pride, or hubris. They saw themselves as the great creators, not God, and the human-built world as evidence of their masterful creative powers.

Systems, Controls, and Information

In chapter 4, I discuss Americans' reactions to the large technological systems spreading through the human-built world. During World War II, engineers, scientists, and managers in the belligerent countries invented and developed horrific technological systems of destruction. In the United States, military technology showed its most awesome power culminating in the atomic mushroom cloud that has come to symbolize the deadly forces that technology unleashes. Having once thought of technology as relatively simple, controllable machines, Americans acknowledged and confronted the omnipresence of enormous, complex technological systems defying control. They also realized that they had become the unchallenged world master of technological forces.

The Manhattan Project that produced the atom bombs became the prototype for large postwar technological systems generated by a military-industrial-university complex. I discuss the nature of several of these weapons systems and then turn to the negative reactions of American commentators to this complex.

A search for controls that could harness technology seemingly expanding out of control led engineers and scientists to investigate and write about the interaction of information, communication, and controls, especially feedback controls. In his 1948 book *Cybernetics; or, Control and Communication in the Animal and the Machine,* MIT mathematician Norbert Wiener declares that the seventeenth and early eighteenth centuries were the age of clocks; the later eighteenth and the nineteenth centuries, the age of the steam engines; and the twentieth century, the age of communication and control.

I also consider the reactions of various American commentators to the information revolution, which arose in part because of the search for the means to control large technological systems. Americans' reactions to the information revolution have been as enthusiastic as their early responses to mechanization. George Gilder, an autodidact and public intellectual, has been the most enthusiastic and influential of the celebrants of the computer-driven information age. He declared that the industrial age has passed. Creative human minds are displacing machines and material things as the measure and source of wealth.

Technology and Culture

Chapter 5 examines the works of artists and architects who have helped me understand technological creativity and the human-built world. Experiencing a second industrial revolution in Germany and the United States during the late nineteenth and early twentieth centuries, many of them believed that technology was making a modern technology-based culture that would extend into a limitless future. Their architecture and works of art filled the human-built world.

As a means of construction and as a source of symbols, technology shaped an International Style of architecture that expressed the technological values of rationality, order, and efficiency. Artists turned to technological symbols and metaphors to represent a modern world and human characteristics. Technological values associated with machines infused art and architecture to such an extent that the modern Western culture flourishing between the world wars can be characterized as a technological, or technology-based, culture.

Artists' and architects' reactions to technology changed during the twentieth century. Early in the century, machines fascinated many German and American artists and architects. Some even celebrated the machine. After World War II, during the era of systems and controls, leading American artists and some architects reacted strongly against technology, seeing its order and control as oppressively confining the human spirit. As the century closed and an information revolution spread, some American artists and architects began to view technology positively again, seeing computers as a creative tool.

Ecotechnological Environment

In chapter 6, we turn from my sources to reflect upon the accelerating spread of the human-built world, an overarching theme in the preceding chapters. I will consider our moral responsibilities as creators who have embodied our values in the technological systems constituting this world. These values react back upon us and shape our behavior. Because of their physicality and durability, technological systems pass on the virtues and sins of fathers and mothers to sons and daughters.

While environmentalists are deeply and rightly concerned about the natural world, we should also seek ways in which we can make the human-built world more sustainable and supportive of the quality of human life. Yet we should not focus simply on the human-built, but direct attention to the ecotechnological environment in which we mostly live, systems in which the natural and human-built overlap and interact.

I also reflect upon the ways in which Americans' infatuation with technology had led them to favor technological fixes, which are efforts to solve complex problems with relatively simple technology. Consequentially, the problems are usually not solved, but often made more intractable. For instance, the American government in recent years has been inclined to use overwhelming military technology to solve long-standing and complex political, economic, and social problems throughout the globe.

Finally, I ask if increasing the technological literacy of Americans and their participation in decision making involving a large measure of technology will improve the ecotechnological environment and reduce dependency on technological fixes.

Technology and the Second Creation

Today we rarely attempt to understand technology by placing it in a religious context. In the late nineteenth century, the evolutionary science of Charles Darwin persuaded open-minded Western people that a religious perspective did not help them understand science, much less applied science, or technology. They understood technology as materialistic and practical, requiring a commonsensical approach in order to use it. Engineers, mechanics, and artisans practiced technology, and they did not need a philosophical or theological grounding to excel at their tasks.

Earlier, however, literate men and women in Europe and America contemplating technology's transformation of a natural into a human-built world situated technology as a creative tool into a religious context. Many persons with a theological or philosophical turn of mind who thought about technology believed it to be a God-given way in which to recover a lost paradise, or Edenic state.

The concept of a second creation and the Edenic recovery has a long history. Cicero in *De natura deorum* (45 B.C.) conceived of a second nature that humans create by channeling the rivers to suit their needs, by sowing and fertilizing the soil of the plains and the mountains to bear fruit and wheat, and by planting trees to shade their gardens and parks. Medieval Catholic and later Protestant theologians believed that Christians possessed a divine creative spark that would enable them to design tools and machines capable of transforming the land into a new Garden of Eden.

Mephistophelean Creativity

The second creation theme reaches a peak in nineteenth-century America, but the theme has a long history in Western religious and philosophical thought. German poet, statesman, and scientist Johann Wolfgang von Goethe vividly expressed the creative drive of his contemporaries in his epic poem *Faust*. Goethe's presentation of the hoary Faust myth differs from other versions in its emphasis upon demonic creativity. Human creativity for Goethe became the mark of a divine life. At the same time, Goethe found creative humans, such as Faust, guilty of the sin of pride, which, from a theological perspective, is the worst of the seven deadly sins. A prideful human, like the fallen angel Mephistopheles, challenges the almighty power of God as the Creator. Goethe's *Faust* allows us to see the egotistical and controlling nether side of creativity.

Completed in 1832, *Faust* presents subtle insights into the second creation, or the technological transformation of the material world. Goethe plumbed the depths of human creativity, finding there both the benign and the wicked. In the poem's opening, we find Faust, an aging scholar, waging his eternal soul that Mephistopheles, a fallen angel who involves himself in human affairs, will be unable to provide him with an earthly experience sufficiently engaging to cause him to abandon his ceaseless striving and to rest content with an earthly achievement. He dares Mephistopheles to provide food that does not satiate, gold that does not pass through his fingers, and a woman who does not lust for his neighbor. Undaunted, the fallen angel promises Faust experiences that will cause him to ask the moment to linger.

The wager made, Mephistopheles conjures up for Faust one earthly pleasure after another—intellectual, sensuous, and sensual—without Faust's finding fulfillment in such fleeting moments. Only after Mephistopheles helps Faust preside over one of the most symbol-laden, large-scale technological projects in literature is Faust brought to ask that the moment of fulfillment linger. Not knowledge

2. *Because of the godlike and prideful act of parting the waters and creating land, Faust almost loses his soul to Mephistopheles. The angelic Margarete, however, saves him. Paul Struck,* Fausts Ende *(1973). Courtesy of the Goethe-Museum Düsseldorf/Anton-und-Katharina-Kippenberg-Stiftung.*

of the innermost workings of the world, not the love of the maid Margarete, not union with Helen of Troy, not the fleshpots of Walpurgis night, but the prospect of a land-reclamation scheme— of nature transformed—almost causes him to lose his spirit to Mephistopheles. Creating land by parting the waters in the manner of God in the book of Genesis, Faust exults in the creation of order from chaos and the making of a new Eden populated by a community of free men and women.

Goethe also explored the nether side of creativity by having Faust exhibit a neurotic craving to control land and people. He is irritated beyond reason to discover that an elderly couple, Philemon and Baucis, have reclaimed and settled a small plot of land long before his arrival and that they doggedly resist the "developers" who would level their cottage and chapel. Faust finds this defiance, this exception to his masterpiece, intolerable. Frustrated in the un-bounded exercise of his godlike will by a patch of land still outside his control, he allows Mephistopheles to violently remove the old couple, who die from the shock. Like some other technological sys-tem builders, Faust also exposes the morally questionable side of his project by having Mephistopheles use diabolical means to pro-vide thousands of slavelike workers to build the dikes and drainage canals that will reclaim the land from the sea.

Goethe uncannily anticipated the behavior of twentieth-century system builders engaged in totalitarian projects. Motivated by Marxist moral arrogance and empowered by authoritarianism, So-viet engineers and managers with an American-style affection for megaprojects ran roughshod over workers and local communities as they built canals, railroads, hydroelectric projects, and industrial complexes. Construction workers often survived miserably in tents and mud huts surrounded by open sewers. The technocratic poli-ticians insisted that the projects prepared the way for a socialist utopia. The Soviet style of system building has spread recently to China. The Three Gorges Hydroelectric Project on the Yangtze River now under construction may displace more than a million people and flood more than 100 towns, 800 villages, and almost 100,000 hectares of prime farmland.

Mechanical Arts

Goethe's conviction that humans exuberantly express themselves in second creation projects involving the domination of nature was deeply rooted. The instinct to transform the world in a godlike manner can be traced back at least to medieval and Renaissance culture. In sharp contrast to believing a fallen angel to be their cre-ative ally in questionable pursuits, medieval Catholic and later Prot-estant theologians imagined Christians fulfilling divine purpose as they mastered the earth. They reasoned that possessing a divine cre-ative spark, humans could design tools and machines that would allow them to transform the land into a new Garden of Eden.

The Benedictine order of monks stressed the importance of manual landscape-altering work along with meditation and liturgical praise. The reforming Cistercian order became known in the twelfth century for its development of mechanical devices such as water mills and windmills that ground grain, sawed wood, worked metals, and reclaimed the wetlands.

The success of second creation work depended upon the invention and use of tools and machines. John Scotus Erigena (ca. 810 – ca. 877), who flourished at the Carolingian court, and Hugo of St. Victor, the twelfth-century director of the School of St. Victor in Paris, celebrated machine technology by discussing the mechanical arts and raising them to the status of the liberal arts so celebrated in antiquity. They argued that humans should cultivate the mechanical arts of weaving, weapon making, navigation, agriculture, hunting, medicine, and drama so that they would flourish alongside the liberal ones of grammar, dialectic, rhetoric, geometry, arithmetic, astronomy, and harmony.

In handsomely illustrated sixteenth-century books describing military and civil machines, architects and engineers continued the celebration of the mechanical arts. They argued, like Hugo of St. Victor before them, that the mechanical arts offered a way to recover the Edenic paradise. They believed that over the millennia humans have gradually recovered reason lost with the ouster from Eden, and that with reason the competence for mathematical thought and, therefore, for machine design and construction would develop.

As instruments embodying reason, machines could be used to create an environment equal to, or even surpassing, the one enjoyed before the Fall. Furthermore, several of the authors believed that designing and using machines for such ends is far more spirit fulfilling than lounging about in Eden as Adam and Eve did before their ouster from the Garden. In essence, machine book authors assumed that the ability to develop machines is God-given and that

3. Medieval monks elevated the mechanical, or technological, arts to the level of the liberal arts. Portrayed in cathedral stone, a perpetual motion mill grinds grain. Le Moulin mystique (The Mystic Mill). Nave capital. Ste. Madeleine, Vezelay, France. © *Foto Marburg / Art Resource, NY.*

machines are wondrous means of satisfying the human longing to create.

A sixteenth-century German physician, alchemist, and chemist, Theophrastus Bombast von Hohenheim, or Paracelsus, also stressed that the mechanical arts make possible godlike creativity. His version of the second creation theme has God leaving nature unfinished and its essential usefulness concealed in dross, thus providing a creative challenge for humans. According to Paracelsus, before the Fall and ouster from Eden, humans did not need the mechanical arts, and they did not have to work in order to transform the landscape. After the Fall, their circumstances changed and became like ours today. Even though a merciful God allowed the angels to impart their knowledge of the mechanical arts to Adam and Eve, they could only survive through work, discovery, and invention. Humans, henceforth, would eat their bread in the sweat of their brow as they engaged in a second creation. In the wilderness into which they were thrust, God had concealed the secrets of natural forces and the materials that humans needed. These they had to ferret out by reason and craft.

Technological Millennium

In the sixteenth and the seventeenth centuries, Puritanism spread from Geneva, where Protestant reformer John Calvin governed a theocracy through northern Europe, England, and Scotland. A reform movement, Puritanism rejected the primacy of the pope, called for the translation of the Latin Bible into German and English, rejected the celibacy of the clergy, and instituted an austere church architecture. Puritanism also spread a Protestant work ethic, or set of values, that continues to influence Western attitudes toward technology. Calvin preached that worldly success in trade, manufacturing, and mining indicated that a person was among the elect chosen for salvation by the Almighty. An individual's commitment

4. Ingenious sixteenth-century German craftsmen designed and constructed a water-driven "two-directional hoist" for lifting ore from deep mines. Illustration from Georgius Agricola, De re metallica *(1557). Courtesy of the Burndy Library, Dibner Institute for the History of Science and Technology, Cambridge, Massachusetts.*

5. After the ouster from Eden, Eve and Adam had to labor by the sweat of their brow with technological tools. Andrea Pisano, The First Labor After the Fall: Eve Spinning, Adam Delving. *Courtesy of the Museo dell'Opera di Santa Maria del Fiore—Firenze. Photograph by Nicolò Orsi Battaglini.*

to work likewise suggested election. The work ethic reinforced a belief that work and technology would restore the Edenic state.

Calvinists and like-minded Anglicans, Presbyterians, and Quak-
ers in England, Scotland, and colonial America believed that natural philosophy, or natural science, as well as mechanical arts and
work provided a rational and empirical means to improve their
earthly condition, or material circumstances. Francis Bacon, an
influential English essayist and statesman, articulated a philosophy
of science and technology that drew upon earlier Christian thought
and that substantially molded the work ethic and the Puritan world-
view. In *Advancement of Learning* (1605), he wrote that scientific activity has as its appropriate end the glory of the Creator and the improvement of man's estate. In his *Novum Organum* (1620), he foresaw
technological knowledge, especially as embodied in machines, as a
way to recover from the Edenic Fall and to regain a paradise characterized as within the reach of all men.

Bacon urged elite men of philosophy to converse with craftsmen
in order to unite natural philosophy and the mechanical arts. Philosophers could rationalize craft practice, and craftsmen could
bring the philosophers face-to-face with the physical world that they
sought to understand. In his essay *New Atlantis* (1627), Bacon imagined wise men in a house of philosophy (Solomon's House) applying
philosophy to the mechanical arts. They and other Puritans of a
New Jerusalem would have dominion over nature.

Bacon perceived voyages of discovery, developments in printing, metallurgy, optics, and ballistics, the rise of Protestantism, and
the general increase of knowledge as evidence that the millennium,
the New Jerusalem, was close at hand. He anticipated a dawning
day when God and man again would become coworkers in creation. In preparation for the coming of the New Jerusalem, Puritans
following Bacon dedicated themselves to the pursuit of practical scientific knowledge and technology, values that did then and do now
characterize Western culture.

A poet, John Milton, author of *Paradise Lost* (1667), stressed the
nature-domination theme, too. Humans armed with the mechani-

cal arts and science would have dominion over the earth, seas, and heaven. Mother Nature would surrender to man. Characterization of nature as feminine became an oft-sounded theme among second creators in the Old and New Worlds.

Other influential Puritan thinkers predicted that a great Puritan-inspired instauration, or restoration after social decay, would bring on the millennium. In their millenarian faith, the Puritans foresaw an early victory over the Antichrist and the rule of the saints in a New Jerusalem not unlike the old Eden. Those who colonized the New World, especially the Puritans of New England, believed that the second creation of the New World would bring the millennium. Even today many Americans see themselves as the world's most technologically creative people who are presiding over the making of a promised land.

Creating a Promised Land

Early settlers in colonial America brought with them northern European attitudes, especially Puritan beliefs, toward work, technology, the second creation, and the Edenic recovery. Some saw their goal as a release from Old World political and religious persecution, and many others longed for an improvement in their material circumstances. A spirit of adventure motivated still others. Like their European and Puritan forebears, many settlers also found the meaning of their great migration as a transformation of a wilderness into a new Eden. To this end, they embarked on deforestation, swamp draining, land reclamation, and other land-transforming work. Conquering what they took to be a wilderness provided Americans — as Brooke Hindle, a historian of early American technology, writes in *Technology in Early America* (1966) — an opportunity to know the thrill of the technological transformation.

In his *History of Plimoth Plantation,* William Bradford, one of the original Puritan colonists, characterized his venture into the New World as a struggle with wilderness. When he and his company came ashore in winter, in a country known to have "cruell & feirce stormes . . . what could they see but a hidious & desolate wildernes, full of wild beasts & wild men? And what multituds ther might be of them they knew not."[1] After corn was harvested, cabins built, and a stockade raised, the Pilgrims survived their first year with the help of friendly Indians. The Pilgrims then celebrated their triumph over nature. They were initiating an American myth of the Edenic recovery.

The magnitude of uncultivated land and dependence on hand tools, horses, simple machines, wind, water, and muscle power, however, circumscribed the first settlers' transformation of nature. In the seventeenth century, many of the colonists, especially the followers of Puritan leader John Winthrop, defined their primary errand as religious and political, as the establishment of a religiously grounded civil government. Following him, they wanted to establish "a Citty upon a Hill" that would be emulated by the corrupt Old World.[2] The city on the hill would become, they believed, a moral beacon.

Later in the eighteenth century, many of the colonists dedicated themselves to the invention and development of a political constitution for a new nation. This done, their errand in the wilderness changed from the founding of the moral beacon on the hill and a new republic to an intensive creation of a living space and workplace for the ever-increasing number of immigrants.

Technological Transformation

Perry Miller, Harvard history professor, literary critic, and a doyen of American studies scholars, has written eloquently about American attitudes and ideas among the reflective classes toward technol-

1. *Bradford's History "Of Plimoth Plantation"* (Boston: Wright & Potter, 1899), p. 95.

2. Quoted in Perry Miller, "Errand into the Wilderness," in *Errand into the Wilderness,* ed. Perry Miller (Cambridge, Mass.: Belknap Press, 1956), p. 11.

ogy and nature. His writings have conclusively demonstrated for me that the history of technology need not be a dull story of nuts and bolts. In *The Life of the Mind in America: From the Revolution to the Civil War* (1965), he describes the general transformation from the eighteenth to the nineteenth centuries in America, especially New England, of attitudes toward science, technology, and wilderness. In the eighteenth century, those who formulated and articulated attitudes—editorial writers, speakers on public occasions, authors of books and articles for educated general readers, and clergy—celebrated the passive contemplation of divine creativity revealed in nature.

The coming of steam engines, steamboats, canals, and railroads in the early nineteenth century changed the ideological landscape from passive contemplation to active transformation. A nation of immigrants, impoverished in their homelands, now empowered by technology, busied themselves improving their worldly circumstances. Articulate spokespersons transformed the settlers' venture into a moral imperative framed in a religious context brought from the Old World. An attitudinal sea change occurred. From the pulpit, at political rallies, in commencement addresses, and in their newspapers and journals, Americans exuberantly celebrated the machines that made possible their second creation of the world. The Faustian will to create had ousted contemplation. Americans would become the lords of creation. Human design was supplementing the Creator's plan for the universe.

Americans in the nineteenth century conceived of the transformations wrought by technology as a manifestation of mind over matter. Rejecting thought that had turned to dry and barren pursuits, they celebrated a new utilitarian age. They believed that creative minds inventing and controlling machines could transform nature by organizing it for human ends. Arrogantly and naively, they boasted that in the nineteenth century the products of the human mind surpassed those of all previous history in grandeur and brilliance. Full of self-satisfaction, they believed that the human mind was ordering chaotic nature, a wilderness, into a world of en-

6. *Early settlers in New England viewed the New World as a howling wilderness. Thomas Cole,* Tornado *(1835), oil on canvas, 46³⁄₈ × 64⁵⁄₈ inches (77.12). In the Collection of The Corcoran Gallery of Art, Museum Purchase, Gallery Fund.*

lightened culture. They chose to ignore the ways in which millions of Native Americans had already transformed "chaotic nature."

Nature as Wilderness

Americans, especially those in New England, heard from the pulpit and podium that the Creator of the universe had instilled within them a creative spark that enabled them to create order from the lifeless elements of nature, a nature made up of unorganized matter and inhabited by inferior animals. They ardently celebrated their domination of nature with words like those found in the biblical book of Isaiah in which the faithful are instructed to make a pathway for their Lord by filling the valleys and lowering hills. The pages of the *North American Review* in 1831 asserted that "mechanism" now provided canals where the Creator failed to make rivers, rolled the rough planet smooth, breached mountains, and propelled vessels across the waters.

In 1814 a commentator complacently observed that America, which two centuries earlier had been a country of wild beasts and savages contending for barren lands, had been made over into a nation of cities cultivating commerce, religion, art, and letters. Another observer imagined a wilderness inhabited by savages and terrible reptiles being marvelously improved by technology. Technological enthusiasts often cast nature as a female object awaiting transformation by men. "Eve" became code for virgin land to be harvested and as fallen nature to be redeemed. Resources are taken from nature's bosom and technology penetrates her womb.

The Native Americans' manifold transformations of nature did not figure in settlers' rhetoric. More than a million Native Americans who occupied the land did not perceive it as wilderness. They had used technology to gather food, cultivate crops, weave materials, build settlements, and generally to create a human-built world. Chief Luther Standing Bear recalled, furthermore, that his people did not imagine the uncultivated plains, forested hills, and winding streams as wild. White men, not Native Americans, saw nature as a wilderness infested with wild animals and savage people. For Chief Standing Bear, nature was bountifully life supporting.

Pastoral Vision: Jefferson

Not all Americans saw themselves as mastering the mechanical arts in order to aggressively transform a wilderness. Not as challenged by a cold, hostile uncultivated land as New Englanders, settlers in Virginia and elsewhere in the South plowed and harvested rich soils during bucolic springs, hot and humid summers, and extended colorful falls. In his 1584 voyage to the New World, Captain Arthur Barlowe described Virginia as a marvelously fertile garden. Not averse to attracting new settlers, Barlowe's reaction differed radically from that of William Bradford, who saw a howling Massachu-setts wilderness. Robert Beverley in his *History and Present State of Virginia* (1705) also portrayed Virginia as nature's garden rich with fruits and flowers and filled with songbirds. Virginia was seen as an Eden inhabited by noble savages.

Thomas Jefferson, a Virginian, preferred that Americans use simple tools to establish a pastoral landscape blending uncultivated nature and a cultivated garden. His poetic vision reveals that the best of intentions eloquently argued often fail to cope with reality. His vision was a variation on the age-old dream of using technology to recover the Edenic state. He knew the writings of Virgil, who contrasted a tranquil pastoral landscape, a green valley, with Rome's troubled urban civilization and with an uncultivated wilderness. Virgil envisioned a middle pastoral landscape that mediated between an uncultivated primitive nature and a complex human-built world.

Jefferson wanted the new nation to become well-ordered gardenlike acres farmed by sturdy husbandmen using simple technology, such as a well-designed plow, which he invented, and water-driven grain mills. Such a pastoral environment, he was convinced, would shape the character of Americans in ways quite different from those that were shaping the denizens of Britain's industrializing urban cities. Dark satanic mills would remain in Europe.

In his *Notes on the State of Virginia* (1785), Jefferson declared that a pastoral landscape peopled by sturdy husbandmen and their industrious wives would nurture genuine virtue. If there were a chosen people of God, they would be these tillers of the soil. He also ventured that history revealed no instance of moral corruption among cultivators of the land. Their goal is the happiness attendant upon self-sufficiency rather than upon material accumulation. By contrast, persons involved in trade and commerce become dependent upon others, subservient, and venal. And the mobs in great cities he compared to impairing sores on the human body.

Over time, increasing technological resources, however, transformed Jefferson's vision of a pastoral garden. He no longer totally rejected the complex machine. He saw the steam engine as a creative tool, but—and this is a substantial "but"—he envisioned machine-driven mills on the countryside, not in dank, dark urban centers like those in the Black Country of England. In his later writings, Jefferson optimistically held that the rural character of America would inculcate in mill owners the same love and care for the land that Jefferson had earlier found in the breast of the sturdy American farmer.

Jefferson and others, as historian Leo Marx has shown, wanted the new nation to situate machines, or technology, in a garden. Marx not only creates a powerful image, but through his book *The Machine in the Garden* (1964) argues that reactions to machines, mills, locomotives, and canals constituted a central concern of leading American literary figures, such as Ralph Waldo Emerson, Mark Twain, and Herman Melville. He suggests that they reflected a more general societal reaction. Marx brings technology to center stage in the drama of American history.

Artist George Inness captured an image of a machine in the garden in his painting *The Lackawanna Valley* (ca. 1856). A striking representation of the idea that technology can blend with nature in a harmonious landscape, the painting has hills and trees gently embracing the industrial buildings and the railroad. No sharp lines divide the man-made from nature. In the 1820s Lowell Mills on the Merrimack River north of Boston also presented, despite their size, an image of a machine in the garden. Hiring young farmwomen, housing them in well-managed boardinghouses, and encouraging their interests in literature and the arts, the mill owners experimented with a new American way of manufacturing.

Paradise Regained

Some Americans had a more radical vision of a technological transformation than simply a middle landscape or a machine in the garden. They foresaw "the Paradise within the Reach of all Men, without Labour, by the Powers of Nature and Machinery." Such was the title of a pamphlet published in 1836 by J. A. Etzler, an American

7. *The locomotive as machine blends harmoniously with nature in an idealized vision of America's exceptionalist encounter with technology.* George Inness, The Lackawanna Valley *(ca. 1856), oil on canvas, 86 × 127.5 cm (33⁷⁄₈ × 50³⁄₁₆ inches). Gift of Mrs. Huttleston Rogers. Photograph © 2002 Board of Trustees, National Gallery of Art, Washington, DC.*

8. *Lowell Mills as a machine in the garden.* East View of Lowell, Massachusetts, *drawn by J. W. Barber and engraved by E. L. Barber. Courtesy of the Center for Lowell History, Lowell, Massachusetts.*

of German origin. He intended to use high technology, figuratively and literally, to bring about the new Eden. Etzler's utopian dream fueled by a myopic view of the role of technology in human affairs suggests the long history of naive technological enthusiasm among Americans.

Etzler provided a yeasty mix of dominion-over-nature aspirations and utopian technological-transformation expectations. He expounded a myth of human creativity rivaling that of the Almighty first mover. His is a materialistic vision recalling, but differing from, the more subtle and perverse one found in Goethe's *Faust*. Etzler spoke for those who yearned for the chance to yield to the gratifications of technology.

He asked why humanity had not used power drawn from sun, wind, falling water, tides, and steam to become the lord of creation. Offering as an example the potential power in the wind, Etzler asked his readers to imagine sails 200 feet high extending over a mile and having 1 million square feet of surface. Assuming a device that would turn them into the wind, he declared that his sails would develop power equal to 1 horse every 100 square feet, so his arrangement of sails would generate 10,000 horses' power. Since men can only work half the day, his sails would provide power equal to 200,000 men. If required, the application of this power could be extended into the clouds by means of kites. Etzler estimated that if the surface of the globe were provided with land sails of his design, they would provide 80,000 times as much work as all the men on earth. The tides, he added, offer the additional power of 2.5 billion men.

Among machines, he foresaw agricultural combines that reap, thresh, grind, and even plow. These could convert the most hideous and sterile wilderness into a bountiful and delightful garden. Machines could transform the post-Edenic wilderness into a pastoral paradise surpassing in beauty and utility even the Garden of Adam and Eve. If a garden were not enough to entrance his readers and move them into a world-transforming technological torrent, Etzler promised that humans, spared labor by machines, would live in magnificent palaces surrounded by luxury and lush gardens.

What an intoxicating vision for poor settlers! And all of this might be achieved in one year without exhausting human labor as mountains were leveled, valleys were filled, lakes created, and the land filled with canals and roads to be used by mechanically propelled vehicles traveling forty miles per hour. In this second creation environment, Americans, Etzler assured them, would lead a life of continual happiness.

Henry David Thoreau, the naturalist, transcendentalist, and great friend of nature, was not quite so sure. In 1843 he reviewed Etzler's *Paradise within the Reach of All Men*, in an essay entitled "Paradise (to Be) Regained." Preferring the first creation to the second, Thoreau believed that the dappled sunlight falling across the path of the poet provokes joy beyond that which human technology can bring. The gentle wind cooling the heated brow fills the poetic mind with profit and happiness equal to that which inventions supply.

Thoreau felt that the enriched and contemplative mind would nurture the human spirit and convey a sense of well-being more fulfilling than machines that tame the wilderness and its creatures. Thoreau did not disparage the inferior energy deities of wind, wave, tide, and sunshine paid homage by Etzler, but ventured metaphorically that the moral horsepower of the transforming force of love exceeds the mechanical horsepower of machines. Thoreau doubted Etzler's prediction that humans would rationally cooperate to rapidly bring about a radical second creation of the world. Thoreau believed that at best the existing one can be only modestly and slowly modified. Yet Etzler, not Thoreau, spoke for his times. Americans then panted for the technological future.

Thoreau identified, if only in passing, a fundamental flaw in visionary utopias such as Etzler's. In all probability, steam, nuclear, internal combustion, and electric power available today exceeds that which Etzler believed could be obtained from wind and other natural sources. Yet the human-built world is not a paradise surpassing in beauty and utility even the Garden of Eden. Thoreau suggested that the inability of humans to rationally cooperate is a cause of this failure. Today we are more likely to attribute the inability to transform the human-built world into an earthly paradise to negative political and social values and structures, than to a failure in rational cooperation.

The values and structures prevailing in the industrialized world have proven, as Thoreau anticipated, highly resistant to change. Consequently, persons with Etzler-like technological visions have tended to become authoritarian masters. George Orwell in *1984* (1949) memorably describes a fictional paradise turned hellish as totalitarian authorities forced humans into "rational cooperation" in order to fulfill a utopian vision. In real life, Vladimir Lenin and his Bolshevik compatriots imposed a totalitarian regime upon Russia after 1917, declaring that such a regime would transform the country through modern technology into a communist paradise.

Emerson: Value-Laden Technology

Ralph Waldo Emerson (1803–1882), the "sage of Concord," Massachusetts, celebrated the transforming power of technology. Emerson believed that the nation could express virtuous values as it created a human-built world. The technological transformation of nature gave clear evidence for Emerson of the mastery of mind over matter. An enormously influential writer and speaker, he believed that just as nature manifests the logos, or the word of God, nature transformed by humans expresses the perpetual creativity and imprint of the human mind. Ultimately the world will be organized to reflect human mind and will.

Emerson realized that artifacts and machines were value laden. Imprinted by the mind, the human-built world reveals human values. Emerson also wanted the heart to shape the physical world. A creative mind and a good heart could leave their mark on a ship, mill, railroad, electric battery, and the chemist's retort, which could then serve God's purposes. He explained that mercenary impulses produced the selfish and cruel aspects of mills, railways, and machinery. Infused with love and created through the powers of science and technology, the second creation would bear witness to the creative divine spark in its goodness and glory. In his 1836 book, *Nature*, he contended that the world lay broken because humans had not fulfilled their moral responsibilities as creators.

Having conceived of the human-built second creation as the fruit of creativity and, potentially, as an expression of love, Emerson assumed the mantle of capitalism's poet-philosopher. His ambivalent view of the natural world accommodated both radical environmentalism and a materialism that celebrated the transformation of nature, which generated wealth. As a divine creation, nature, he decided, was intended to offer its resources to man as a raw material so that he might make them useful. But again Emerson equivocated, for he would not accept the utilitarian and the commercial as final ends. The virtuous mind, he was convinced, ultimately seeks to create an aesthetic environment permeated by architecture and art.

Technological Sublime

Historian David Nye in *American Technological Sublime* (1984) tells us that early American technological enthusiasts celebrated a technological sublime analogous to a natural sublime defined by Edmund Burke (1729–1797), a British philosopher and statesman. Rushing streams in deep, dark valleys and violent storms at sea aroused in him passionate feelings of vastness, power, and infinity. Some Americans, such as Thoreau, shared his feeling for sublimity in pristine nature, but for many others the technological sublime of the second creation increasingly displaced the natural sublime as an object of wonder and contemplation. For them, steam locomotives rushing past in the evening hours, steamboats steering majestically along the broad Mississippi, and bridges spanning deep gorges stimulated wondrous, vast, awesome, and terrifying feelings of sublimity.

Some Americans found Lowell, Massachusetts, textile mills nobler in their majesty than mountains or cascades. A harmonious system of manufacture presented a moral spectacle both beautiful and sublime. For many Americans, a systematic and harmonious interaction of machine parts was analogous to a relationship among individuals in an ideal society. Knowing that nature's power and grandeur displayed in her sublime vistas elevates the contemplative mind and enhances its strength and breadth, the author of "Thoughts on the Moral and Other Indirect Influences of Rail-Road" (1832) found that contemplation and perception of the vastness and magnificence of the country's railroads stimulated a sublime reaction as well. Since the early nineteenth century, Americans have dwelt proudly on their sublime technological vistas, as well as their natural ones. They associate the technological sublime with the country's defining identity. Unfortunately, many Americans today find spectacular, energy-laden weaponry sublime.

Technological Enthusiasm

During the nineteenth century, Americans witnessed the transformation of uncultivated land into a living and working place. Native Americans had used technology to negotiate with nature to make nourishing places on an enormous landmass, but much of the land remained essentially a pristine natural creation. By the early nineteenth century, European settlers, often aided by African slaves, mustered sufficient technological tools, some of their own design and others borrowed from the Old World, to make a frontal attack on what many chose to call a wilderness. Soon they became exhilarated by a technological transformation.

By the second quarter of the century, the British industrial revolution, driven by canals, coal, steam, iron, and factory production, was spreading to the new Republic. Americans adapted British technology to suit their abundance of natural resources and to their lack of skilled labor. An American system of production standardized mass-produced products such as revolvers, axes, and clocks. After canals, railways in conjunction with horses provided transportation networks linking natural resources, production, and market. Steam engines, water mills, and windmills provided power sources for New England textile mills and Mid-Atlantic ironworks and coal mines. The electric telegraph established communication and control linkages for transportation and production systems. Canals, railroads, horse-drawn wagon trains, and ships transported people and goods into the West and later the far West, so the transformation of the natural into the human-built continued. Prodigiously wasteful, Americans exploited their seemingly endless supply of natural resources. American inventors, engineers, workers, and business entrepreneurs built a system of production that, by the end of the century, established the United States as the most productive of nations.

The vision of a machine in the garden and a regained pastoral paradise did not, as Leo Marx has shown, prevail. Writing in 1829, Timothy Walker, a young Harvard graduate and lawyer, sensed a rising tide of technological enthusiasm among the privileged classes. They were witnessing mills at Lowell producing countless yards of textiles; the Erie Canal connecting the Midwest with the eastern seaboard; and the Baltimore and Ohio Railroad inaugurating the railway age. Walker and countless others related progress to the full utilization of the machine; they did not attempt to reconcile technology with nature. The machine need not be in a garden.

Walker expressed his technological enthusiasm in answering an essay by Thomas Carlyle, a prominent British historian and social critic. For Carlyle, his age was not heroic, devotional, philosophical, or moral; it was mechanical in thought and deed. As a mechanical age, it used technological means to achieve utilitarian ends. Not only did a mechanistic approach war with nature to dominate the physical world, but it had also seeped into the spiritual and intellectual world. Concerned about the increasing desecration of the physical environment, Carlyle observed in resignation that even though his age could level mountains and make smooth highways, it lived off spoils wrenched destructively from nature.

In response, Walker acknowledged that some liberties had been taken with nature during the process of mechanization. But, he asked, what is harmful about leveling hills, digging canals, bridging valleys, and tunneling mountains to supply what nature has denied? He cataloged the contributions of machines to cultural improvement. Mechanism, he observed approvingly, forces nature to toil for man. Mechanization emancipates the mind. The nation possessing the most laborsaving machinery, he confidently forecast, would make the greatest intellectual progress. He drew an analogy between the Omnipotent mind creating the machinery of the universe and humans instilled with creative powers creating an ideal mechanized world.

Steam, Coal, and Iron

Thoreau, Jefferson, Emerson, and even Walker did not foresee the cataclysmic changes in the physical environment and in social and political life that the spread of steam engines and associated systems would bring. By the late nineteenth century, the steam engine epitomized the technological transformation as it spun in factories, forged in shops, pumped in mines, and propelled ships and locomotives.

Steam, coal, and iron brought the rise of a mining civilization best exemplified by late-nineteenth-century Pittsburgh, Pennsylvania. Located at the confluence of the Allegheny and Monongahela rivers and at the junction of major railroads, Pittsburgh imported iron ore mined in the Mesabi Range of northern Michigan. Richly endowed with coal deposits, the city built coke ovens, iron furnaces, steel mills, and countless factories. Its industry set the night skies ablaze and polluted the days.

With immigrants pouring into the region to work in the mills and factories, the city grew so rapidly and influential industrialists increased production and profit so carelessly that housing, water supply, sewage, and other municipal functions were overwhelmed or simply ignored. Denied light, fresh air, and sanitary conditions, the workers and their families fell prey to a host of diseases. Technology remained a creative tool, but one that stunted and took lives and dominated and despoiled the environment. It brought unantici-pated changes overwhelming any civic and humanitarian social response not only in Pittsburgh, but also in other late-nineteenth-century cities, including Philadelphia, Baltimore, Cleveland, and Chicago, where Americans could no longer envisage technology bringing a machine in the garden.

A mining civilization also spread westward searching for profits from the production of silver, gold, copper, and other minerals. Copper mining in the Butte, Montana, region after 1890 wasted the environment. Toxic vapors from the mills and the smelters de-stroyed vegetation and livestock on farmlands, and the residue from smelting heaped up in giant tailings. Profiting hugely from the op-eration, the Anaconda Copper Mining Company was the world's fifth largest corporation by 1912. After World War II, the company avoided costly litigation by buying wasted land and introducing some smoke-abatement technology. Even today, however, the Butte region remains one of the most devastated in the nation.

Second Creation Gone Awry?

Pittsburgh and Butte are dramatic instances of Americans desecrat-ing the environment rather than creating a machine in the garden and a pastoral landscape. There are countless other examples. We can ask, to what extent should this misadventure be ascribed to Americans' belief that God has charged them to transform the wil-derness, to use technology and their creativity to master and to sub-due nature for their own materialistic ends? According to the Gen-esis creation story, all of physical creation was intended to serve human purpose. Should the fouling of the environment be attrib-uted to a Judeo-Christian injunction?

An eminent historian, Lynn White Jr. in "The Historical Roots of Our Ecologic Crisis" (1967), has argued eloquently that the story of creation in the book of Genesis persuaded—and still persuades many—believing Americans that God has given them dominion over nature. So empowered, Americans have laid waste to nature in their haste to transform resources into consumer goods. As a result, Judeo-Christianity, according to White, bears a huge burden of guilt. We will continue, he insists, to have a worsening ecological crisis until we reject the Judeo-Christian domination axiom.

Yet, as we have seen in this chapter, many Americans considered their errand in the wilderness to be a means to recover the Edenic state. Eden was a garden, not a factory site for the production of goods. Use the mechanical arts, yes, but for the recovery of a gardenlike environment, is a persistent Judeo-Christian message from the Middle Ages through the nineteenth century. White casts religiously conditioned technology in a negative role. On the other hand, using technology to recover the Edenic state is a message entirely appropriate for our ecologically concerned times, as I shall argue in the final chapter, In the meantime, I shall turn to the ways in which many early-twentieth-century Americans, lamenting Pittsburgh, tried to transform technology into a benevolent machine for production that would sustain both political and economic democracy.

CHAPTER THREE

Technology as Machine

As the twentieth century began, municipal authorities, progressive scientific experts, and well-intentioned middle-class reformers began to respond to and dampen down some of the dire effects of the spread of steam, coal, and iron. Relatively clean electric power, especially hydroelectric power, transmitted from central generating stations over large areas relieved some of the early polluting concentration of industry at railway junctions in industrial cities. Electric lighting in the workplace and on the streets, along with appliances in the home, improved working and living conditions.

Newspapers and magazines began to celebrate the technological transformation. Because machine technology produced a mass of goods and services, consumers responded enthusiastically. The public idolized inventor-entrepreneurs, such as Thomas Edison, and managers of mass production, such as Henry Ford. Technological enthusiasts wanted to organize not only technology, but also the nontechnological world mechanistically. They spoke approvingly in metaphorical terms of the machinery of government, the machinery of the body, and the house as a machine to live in. Mechanization took command at the beginning of the twentieth century much like information would become the hyped technology at century's end.

As living and working conditions in industrial cities began to improve, hundreds of thousands of upwardly mobile American immigrants began to see technology as a horn of plenty. In Germany, liberals idealized what they believed to be an American technology/democracy symbiosis. In the 1920s, they attempted to use their political power to deploy technology to serve a broader constituency, not only industry and the upper-middle classes.

Thorstein Veblen, an eccentric and brilliant American economist writing at the turn of the century, envisaged the entire American economy as a production machine—infinitely larger than the machine Ford created to make automobiles at Highland Park and River Rouge—but not unlike these factory machines with their endless flow of production. In the United States and Germany, the widespread drive to mechanize appeared so obvious to contemporaries that they spoke of their times as a "machine age" and their society as a "machine civilization."

Second Industrial Revolution

A second industrial revolution brought an era of mechanization to America and Germany after 1880. Independent inventors and then industrial research laboratories carried on invention, research, development, and innovation. Electricity and the internal combustion engine brought the telephone, electric light and power, wireless radio, the automobile, and the airplane. The age of steam was giving way to an electrical one. New materials included steel, aluminum, plastics, and reinforced concrete. Mass production and scientific management transformed industrial management. Organizational and social changes included the rise of giant industrial corporations, the spread of higher technical and scientific education in universities and colleges, the increased influence of professional experts from engineering, sciences, and social sciences, and rapidly expanding metropolitan centers, especially Berlin and New York.

While focusing upon the second industrial revolution in the United States and Germany, this chapter will occasionally refer to the efforts of the communist leaders of the Soviet Union to transform agrarian Russia into a modern machine-age nation. The masters of the Soviet Union in the 1920s had more faith in the machine than had any other leaders in history. Surprising as it may seem to many Americans who consider communism an alien, incomprehensible system, the Soviet Union's policy makers for decades looked to American technology as a model.

Henry Ford's automobile plants at Highland Park and River Rouge represented for the press and the public the essence of mechanized America. Designed by a renowned American industrial architect, Albert Kahn, the River Rouge buildings expressed the rational, technological values of order and efficiency. The factory complex, like medieval cathedrals earlier, expressed the essential spirit of an age.

Ford plants stood at the core of an enormous system of manufacture that interconnected assembly lines with satellite producers of parts and raw materials throughout the world. Flow and system characterized the Ford production empire. His low-priced Model T automobile for the masses associated production with democracy and social change. German socialists believed that Ford's mode of mass production could become the hallmark of German white socialism and forestall the spread of red, or communist, socialism. Adolf Hitler's National Socialist Party praised Ford as a great entrepreneur of the machine age. Early leaders of the Soviet Union also idealized him.

Thomas Edison inventing like a wizard in his Menlo Park laboratory, electrical engineers bending over their drawing boards in Berlin and Schenectady, New York, and white-coated scientists experimenting in industrial and university laboratories brought the electrification of industry and cities in America and Germany. New York lit a path into the future when Edison initiated centralized electric supply by illuminating Manhattan with incandescent lamps. Because of the presence of two of the world's leading electrical manufacturers and a highly efficient electrical supply system, Berlin became known as the "city of light."

Although electricity is an ethereal force, journalists, public intellectuals, and social scientists simply subsumed the transformation brought by electrification under the rubric of mechanization. They found it easier to conceive and speak of physical things such as steam-driven electrical dynamos than of the transmission and distribution of intangible electric power.

The spread of electrical power led to the mystification of technology. Electricity so fired the imagination of Vladimir Lenin, leader of the newly established Soviet Union, that he predicted in 1918 that electrical engineers constructing a national power grid and the communist party establishing a new political system would usher in a communist utopia, which differed only in detail from the new Garden of Eden envisioned earlier by Americans.

An American historian, Henry Adams, conveyed eloquently, if obliquely, the mystification of contemporaries confronting electric power. In 1900, after seeing a forty-foot-high electric dynamo at a French industrial exhibition, he described the dynamo as a moral force bringing industrial and social change. He compared it to the medieval Virgin who moved Christians to build great cathedrals and create great art. In 1900 the dynamo seemed to Adams to challenge, but not surpass, the Virgin's attractive force. He found himself figuratively praying to the dynamo.

Modern Metropolis

The machines brought by the second industrial revolution were nowhere more obvious than in the mushrooming industrial cities in the United States and Germany. Early in the nineteenth century, some Americans envisioned mechanization transforming uncultivated land into a pastoral landscape. Machine technology would be the means for an Edenic recovery. Early in the twentieth century, they and the Germans saw machine technology as a means to transform industrial cities into handsome financial, commercial, and cultural metropolises. Not a new Eden, but Berlin and New York became the stage upon which to play out the human drama. Electric

9. *Historian Henry Adam believed that the Virgin Mary energized the medieval world as much as the electric dynamo energized modern society. Virgin and child in west portal tympanum, cathedral, Chartres, France.* © *Foto Marburg/Art Resource, NY.*

signboards and electrically lit department-store windows attracted middle-class shoppers clothed in machine-made dresses and suits; telephone networks linked businesses and neighbors; the brightly lit marquees of theaters illuminated the faces of people excitedly seeking diversions; railway stations and subways witnessed the influx of people from the countryside escaping traditional culture and seeking modern novelty.

Steel-framed bridges and buildings, telephone wires, and radio transmission towers rose above a human-built urbanscape. Moving energy, information, water, waste, food, goods, and masses of people, modern power, communication, and transportation systems made possible the life in the modern industrial city. Subways and underground networks along with tall buildings gave the cities three dimensions that allowed people to mass in unprecedented numbers. American artist Louis Lozowick found a drive to order and organization revealed by the rigid geometry of American cities, "in the verticals of its smoke stacks, in the parallels of its car tracks, the squares of its streets, the cubes of its factories, the arc of its bridges, the cylinders of its gas tanks." [1]

By 1900 Berlin, the capital of recently united Germany, more than any other city, embraced essential institutions of modernity. Two of the world's leading electrical manufacturers, capital-rich investment banks, and a great technical and a renowned liberal arts university, impressively strong in science, flourished there. Berlin theater, music, art, and architecture nurtured an influential avant-garde committed to both a machine culture and a traditional high culture. Industrial magnates, world-renowned physicists and chemists, politicians governing the new state, and avant-garde artists and musicians mingled in theaters, palatial homes, and the corridors of power. New York, London, Paris, and Vienna did not encompass such a modern mix of industry, technology, science, politics, and high culture.

1. Louis Lozowick, "The Americanization of Art," *Machine-Age Exposition Catalogue: The Little Review* 12 (1926–29): 18.

Unlike Berlin, turn-of-the-century New York was not influenced by the presence of heavy industry and a nation's capital. Industrialists and nation-building politicians did not mix with social and cultural leaders, as in Berlin. On the other hand, New York's high-rise architecture and transportation systems were more impressive than Berlin's. Electricity made modern New York possible. In 1882 Thomas Edison established in the Wall Street district of Manhattan the world's first central station for supplying incandescent light and power. Electric elevators made high-rise architecture feasible, and electric power made extensive subways possible. Electric trains brought passengers underground into Grand Central and Pennsylvania stations in the heart of the city. Electric motors supplied power to numerous small industries that clustered in New York City.

The rapid growth of Berlin and New York City has been conventionally attributed to events like the unification of Germany in 1871 and to the rise of a national commercial and trading market in the United States. Relatively little attention has been given to the workers, inventors, and system builders who flourished in these cities and who developed the large-scale technological systems, as well as the business, industrial, and residential architecture.

New York could not have become a world metropolis, a center of commerce and trade, if its engineers and system builders had not found the means to connect Manhattan Island to the mainland. During remarkable decades of heroic technological achievement from 1880 to 1910, they constructed bridges over the East River and tunnels under the Hudson. The Brooklyn Bridge, opened in 1883, remains the most famous and the most aesthetically impressive, but the Williamsburg Bridge (completed in 1903), the Queensborough Bridge (1909), and the Manhattan Bridge (1910) also spanned the East River. Railroad engineers and workers laid tunnels under the Hudson River and elevated railroads over, and subways under, Manhattan streets.

After World War I, Berliners reacted to the shortage of housing with much the same resourcefulness as New Yorkers had responded earlier to transportation problems. Martin Wagner, who served as city architect and as municipal planner in Berlin from 1919 to 1932, used new technology, especially American production technology, to help relieve the city's dire housing problem. Deeply committed to solving the pressing social problems of his day, he designed and presided over the construction of housing settlements that brought workers out of dingy housing blocks into light, airy, and sunny apartments. He is best remembered for the Hufeisensiedlung (horseshoe settlement), a multi-unit housing complex that he designed in association with the architect Bruno Taut. They and other architects responded from 1920 to 1933 to the housing problem with a number of model housing settlements.

Blueprinting and Controlling

Many thoughtful Americans and Germans assumed enthusiastically that the machine gave them a means for designing and solving the problems not only in industrialized cities, but also the human-built world in accord with a blueprint representing their interests and needs. Earth-transforming mechanization also heightened their self-esteem, even their arrogance. In the nineteenth century, Americans had created a human-built world believing that the creative act depended upon a God-given spark and that they were completing God's creation. In the twentieth century, God was no longer needed. Germans no longer needed nature. They were confident that their university and industrial laboratories could create substitutes for the natural resources, such as oil, that nature denied them.

Aware of the industrial transformation in the United States, leaders of the newly formed Soviet Union expressed an arrogance, or hubris, even greater than that of the Americans. Longing for industrial revolution, Leon Trotsky, a leader in the new regime, called for engineers and workers to rearrange mountains and rivers to improve greatly on nature. Man, he wrote, will rebuild the earth "according to his own taste."[2] Technological enthusiasts expected the machine to fulfill their visions, much as enthusiasts today envision the computer bringing a longed-for world.

Technological enthusiasts loved machines because they believed that they could control them, a belief stimulated by the ways in which inventors and engineers were exercising subtle and elegant control of machines. An elementary device, for example, consisting of a hand-sized electromagnet combined with a mechanical spring could control the output of a giant electrical generator. Such a simple device controlling massive forces persuaded engineers that they had the legendary lever with which Galileo Galilei envisioned moving the world.

Load-dispatching centers for regional electric power systems offered an even more dramatic example of engineers' ability to control large and complex machines. Two or three men sitting in a soundless control room monitoring illuminated dials remotely manipulated great switches that controlled the volume and location of electricity flows along transmission and distribution lines electrifying an entire region. Such control rooms anticipated post–World War II mission control centers for the astronauts.

Faustian Technology

Technological enthusiasm prevailed in the popular press during the second industrial revolution much as it does during today's information revolution. Yet public intellectuals, social scientists, and historians, especially in Germany, questioned popular technological enthusiasm and cast doubts about the social and cultural impact of technology. Seeing technology through a cultural filter, they

2. Leon Trotsky, *Literature and Revolution* (Ann Arbor: University of Michigan, 1960), pp. 251–53; quoted in Loren Graham, "Adapting to New Technologies," in *Soviet Social Problems*, ed. Anthony Jones, Walter D. Connor, and David E. Powell (Boulder: Westview Press, 1991), p. 299.

viewed the machine ambivalently. For reflective Germans, technology did not create a human-built world out of a wilderness; they considered it as an intervening force, often disruptive, in a built world heavily laden with cultural traditions. Some associated technological creativity with the prideful Faustian bargain of Goethe's land-reclamation project. Germans and Americans who criticized technology during the machine age provide an understanding of the negative, as well as the positive, character of technology.

Oswald Spengler, a German historian, associated machine technology with the decline of Western civilization. In the twentieth century, it was triumphantly mastering nature, but the West was losing its moral and cultural center. In his influential study of world civilizations and cultures entitled *The Decline of the West* (two volumes, 1918–22), he argues that a cultural sea change occurred when humans began using technology to exploit nature. In contrast to nineteenth-century Americans who conceived of the machine as a gift of God and as a means of regaining paradise, Spengler, like Goethe, characterizes technology as a Mephistophelean instrument, not a benign God-given one. Mastering nature with the machine leads to human pride, or hubris. In a human-built world, prideful humans see themselves, not God, as the all-powerful creator.

By characterizing creativity as having a dark satanic side, Spengler recalls the Prometheus myth. Because Prometheus stole the fire of creativity from the gods and bestowed it upon humans, Zeus punished him and humans by creating and sending Pandora to Earth, where she opened a box filled with human misery and hard work. This creation myth resembles the Genesis story of Eve tempting Adam to taste the apple from the tree of knowledge, which can be understood as a source of worldly creativity. Having forbidden them to share his creativity, God casts them both out of Eden into a world of toil and misery.

In *The Decline of the West*, a somber tome cast in the style of German conservative romanticism, Spengler's arguments are both insightful and irrational. He portrays the history of various cultures over the ages, thus anticipating the British historian Arnold Toyn-

10. After World War I, German historian Oswald Spengler attributed to machine technology modern society's loss of culture and its descent into a sterile machine civilization. Photograph of Spengler (1930) © Bildarchiv Preussischer Kulturbesitz, Berlin.

bee, whose multivolume interpretation of world history, *A Study of History* (twelve volumes, 1934–61), takes successive world civilizations, or cultures, as the defining subject of history. Both Spengler and Toynbee attracted a wide readership among the public as well as frequent dismissive scorn from scholars. Some critics find Spengler setting the intellectual stage for National Socialism with his portrayal of a cosmic twentieth-century struggle between "money and blood." The coming of Caesarism, he predicts, will break the *dictature* of money and its political weapon, democracy.

Medieval monks persuaded themselves that in wresting useful

forces and materials from nature, they were fulfilling God's will. In seeking to dominate nature, however, they, according to Spengler, actually gave the devil a hand in the game. Unlike Christ on the mountaintop, they did not reject the devil's offer of dominion over the earthly environment. Cistercian monks used natural forces and machines to transform a wilderness into fertile fields. Their mastery of water mills and other machines gave them, not God, control over a mechanical cosmos. They forced secrets out of God and nature in order to be godlike creators and controllers.

As machines fell into the hands of laymen, they, too, felt a god-like, prideful dominion over the earth. Artisans and engineers created and celebrated machines because they were small cosmos obeying their will. In their struggle to be masters of a mechanized cosmos, humans denied sacred causality. The unpredictable and uncontrolled play of natural forces once reminded humans of their limited power and of forces beyond their control that were shaping their lives. Yet, in the machine age, a child could control fabulous forces seemingly more powerful than God-given natural forces.

Spengler contrasted a past Western culture, with its aesthetic and moral concerns, with the rise of a Western civilization that worshiped production and engineering works. Engineers became the creative cultural heroes of Western civilization, as they presided over the machinery of the second industrial revolution. They flourished as Western history entered its autumnal era. They designed automobiles; trains that crossed continents in a few days; ocean liners that resembled floating cities; airplanes that erased space and time; telegraphs, telephones, and wireless radios that sent messages with the speed of light; buildings that scraped the clouds; and monstrous dynamos controlled by the flip of a switch. Spengler, like a creative engineer, relished the aesthetics, the subtlety and elegance, of well-designed technical artifacts. Steely engineers, for Spengler, had become the priests of the age of mechanization. Like the earlier priests of the church, engineers interpreted and dispensed transcendental powers. They drew upon these to design and energize the machines, which they mastered and controlled.

Yet Spengler raised the possibility that in the future engineers might find irrational spiritual health more important than rationally mastering the forces of nature. Suppose, he proposed, that mysticism should overwhelm their reason. Imagine, he added, that engineers should finally become aware of, and reject, the God-defying nature of the machine. This would bring an end to machine civilization, leaving only the ruins of the Faustian technology that altered the face of the earth.

Demonstrating his continuing interest in technology and its impact, Spengler in 1931 published a slight volume entitled *Man and Technics*. Despite its prescient anticipation of technology's increasing desecration of the environment, the book had a mixed reception. H. Stuart Hughes, a cultural historian, asserts that Spengler's *Man and Technics*, unlike *Decline of the West*, made little impression on the literate world. Hughes finds much of it embarrassing. On the other hand, John Farrenkopf in *Prophet of Decline: Spengler on World History and Politics* (2001) calls *Man and Technics* a gem of a little book in which Spengler addresses the disastrous impact of technology upon the natural environment.

In *Man and Technics*, Spengler denies that Western civilization is deploying technology to bring the greatest happiness to the greatest number. Instead, humans enthusiastically use technology to devastate nature and reign as a deity in its human-built replacement, themes adumbrated in *The Decline of the West*. World history reveals humans as inventive "beasts of prey," exulting as they use machines to "plunder" and "rape" nature. Spengler anticipated an ecological crisis as he observed world mechanization eliminating forests and species, as well as threatening agriculture because of climate change. Spengler's insight is even more notable because his eminent predecessors Immanuel Kant, Georg Wilhelm Friedrich Hegel, Karl Marx, and Max Weber all failed to recognize humans' baleful impact upon nature.

Spengler proved to be less prescient in his prediction that non-Western peoples would displace Westerners as the masters of technology. Because he was witnessing the transfer of technology and engineers to Asia and to Russia, he believed that these regions with their cheaper labor would dominate world trade in the near future. In the late twentieth century, Japan and Pacific Rim nations seemed to be fulfilling his prediction, but the United States has established itself in the twenty-first century as the world's preeminent technological power.

Spengler's belief in the technological prowess of the Faustian, especially German, civilization anticipated National Socialist racism. Hitler in *Mein Kampf* declares that mankind could be divided into three groups: the founders of culture, the bearers of culture, and the destroyers of culture. Aryans, principally Germans, Hitler argues, have created humankind's great works in art, science, and technology; other peoples, including the Japanese, have been culture bearers; and the Jews, culture destroyers. The lower races should, Hitler writes in a darkly prophetic passage, labor for the culture-building Aryans. Hitler portrays Jews as the acquisitive bankers of the Western world. In predicting a titanic struggle between creative engineers and acquisitive financiers to control the productive forces of the world, Spengler fed into Nazi racist ideology.

By insisting that Nazi ideology would reinforce the creative powers of engineers, Nazis responded to Spengler's warning that engineers should not give into irrational mysticism. Nazi engineer Fritz Todt, whom the propaganda machine characterized as the heroic embodiment of Nazi values, assembled young engineers at retreats in historic castles to tell them that cultural values would determine the future of technology and that Nazi culture, as defined by Hitler, would inspire the engineering profession to become supremely achieving. Todt headed large-scale military and civil projects including the Autobahn.

Mechanical versus Organic

Lewis Mumford, an American public intellectual, did not, like Spengler, place technology in a religious frame, but he raised profound questions about the mechanization of society. He believed that the embedding of machine values in a culture threatened to eliminate an organic characteristic needed to sustain a healthy culture that nourishes the human spirit. Because his seminal book about the history of technology, *Technics and Civilization* (1934), draws heavily on Germans, including Werner Sombart and Walther Rathenau, whom we will later consider, he, like them, tempered his technological enthusiasm with skepticism.

Publishing books and articles about technology, Mumford became America's most eloquent and influential twentieth-century critic of technological change. Not a professional academic, he survived as a prolific independent author and public intellectual commenting upon not only technology and society, but upon architecture as well. Over the years, he changed from a sympathetic skeptic to a Jeremiah on matters technological. During the cold war, he portrayed a technology-dependent society as racing downhill toward an abyss in a vehicle without steering gear.

Between the wars, Mumford argued that technology wisely and morally deployed could bring a better world. He predicted that electric power could eliminate many of the grim and grimy landscapes of the steam, coal, and iron era. He envisioned thermal and water-driven power plants located far from cities transmitting clean power to green manufacturing sites spread across the countryside. And he confidently believed that the essentials of the second industrial revolution—electric power, the automobile, radio, and telephone—could transform an industrial society of congested cities into one of economically and demographically balanced regions. No longer would the industrial and commercial population heap up at the mines, at railroad lines along valley bottoms, and at ocean ports, but would spread, as did electric power-transmission lines, to upland and rural areas. This bright future depended, he was persuaded, upon control of technology passing from industrial capitalists to well-informed and well-meaning scientists and social scientists. He did not explain, however, the way in which this social revolution would be brought about.

11. Lewis Mumford, an eloquent and skeptical critic of technology and society, longed for a green life-enhancing technology, but feared that machine technology would deaden the organic in nature and life. Photograph of Mumford courtesy of the Annenberg Rare Book & Manuscript Library, University of Pennsylvania.

Although Mumford anticipated a bright future if the technology of the second industrial revolution were benignly deployed, he warned that unrestrained mechanization would deaden the vital organic aspects of life. His views on the mechanization/organism conflict resemble those of a number of earlier critics of mechanization, including Thomas Carlyle, Matthew Arnold, John Ruskin, and Samuel Taylor Coleridge. They expressed a romantic reaction against the scientific materialism and machine idolatry stemming from the eighteenth-century Enlightenment. Coleridge believed that humans impose mechanical forms on matter. On the other hand, organic forms arise from within; they evolve from innate properties. Nature is the infinitely diverse creator of organic forms. Some architects, namely, Frank Lloyd Wright and Louis Kahn in the twentieth century, have designed organic buildings that they believe evolved from their site or out of their constituent material.

Mumford argued that the process of organic development in nature should be a model not only for architecture, but also for many forms in the human-built world. Humans, he added, should not limit themselves to the creation of machinelike forms, but, like nature, should create organic forms. Mumford, like his predecessors, saw the machine crushing the organic spirit and the emotional and spiritual values critical for sustaining a life of spontaneity and individuality, as contrasted to one of rote rigidity. He believed that interwar machine-worshiping America had become especially prone to losing contact with organic functions and human purposes.

Mumford's insistence that humans need to relate to the organic reverberates sympathetically with the views of those who seek to commune with the organic, natural world. Because most inhabitants of the industrialized world today live face-to-face with the human-built, they encounter things that express the values of other human beings. Nature speaks in a different voice, one that Mumford called organic. Once it spoke to many as a divine creation; today it addresses many more as an environment free from negative human impact.

Transcending the Organic

In contrast to Mumford, Werner Sombart argued that the machine embodies a prevailing trend of the modern age, which is to free humans from the limitations of the organic. A German economist, historian, and Berlin professor with a more substantial reputation among scholars than his contemporary Spengler, Sombart wrote a massive three-volume work on capitalism, the title of which trans-

12. German economic historian Werner Sombart realized that technology interacted with other values and forces to create a modern culture. Photograph of Sombart (1935) © Bildarchiv Preussischer Kulturbesitz, Berlin.

lates as *Modern Capitalism: A Historical-Systematic Presentation of European Economic Life from the Beginnings to the Present* (1919). Sombart called the period from about 1750 to 1914 the age of high capitalism, or the machine age.

Machine technology, Sombart concluded, empowers humans to transcend natural limitations—their own and those of the environment. If western Europe had continued to depend upon organic resources, such as wood, natural fertilizers, and draft animals, then the level of population would have remained about the same as it had been in the later Middle Ages. If the German navy had had to rely on wood for construction, instead of iron and steel, all the forests of Germany could not have supplied the materials needed to build the World War I German fleet.

The substitution of coal for wood as a fuel in conjunction with the use of the steam engine provided mechanical horsepower far beyond the highest imaginable level of animal power. Steam engine–driven and electrically driven machine tools performed work inconceivable to a tool-wielding human of a century earlier. Sombart prophesied that humans would design automated machine systems and thus could dispense with workers as machine tenders. Humans would replace a natural world with a machine-ridden world of inorganic matter.

Sombart defined technology as systems that are means to fulfill determined ends. He identified production technology as essential for sustaining modern culture, the technology on which many other technologies depended. He argued that science-based technology, rather than an empirical approach, characterized the modern "style." Yet he had an ambivalent attitude toward technology and modernity, which technology sustained. Like a number of other German social critics, he romanticized a German past rooted in the German forest and soil, a past that cultivated heroic human attributes such as honor and duty rather than reason. He especially scorned the values embodied in rational capitalism and the "trader's" mentality, an attitude of Sombart's that some of his contemporaries considered anti-Semitic.

An anti-Marxist in later life, he denied that the forces of production determined economic and cultural life, believing instead that cultural values could shape the economy and technology. In Sombart's view, cultures, like that of the Romans, could decline while their basic technology remained the same. Two regions, such as modern Chicago and Berlin, have dramatically different political systems but have the same technology of production. Therefore, technology and culture are not bound to one another.

In an essay entitled "The Influence of Technical Inventions," Sombart did discuss at length the ways in which technology often shapes history. He stressed, however, that the shaping was not uni-directional. Other cultural forces shaped technical invention. He wrote of the interaction of capitalism and technology. Investors congregate at the sites of technical invention, and technical invention enhances their wealth and nurtures their capitalistic spirit. He also saw the long-standing interaction of the manager's talent for organization and the inventor's need for organization in order to develop his invention. Sombart perceived and emphasized interactions, or feedback.

Sombart found that the modern state of mind is a function of technical innovation and capitalism. They inculcate quantification and rationalism, which feed back upon capitalistic and inventive activity. Moderns are far more interested in progress as manifest in electrification and wireless telegraphy than in original sin or the thoughts of Goethe. Modern men and women, according to Sombart, have relegated idealistic tendencies to the background while placing the acquisition of material possessions to the fore. He compared modern humanity to cattle grazing in the meadow. Their keepers are the profit-chasing masters of production processes. He scoffed at the acquisitive Americans with their childish smug satisfaction in technical progress and world-leading production. Sombart observed that Americans are more interested in the linotype machine than in the content of the newspapers it prints.

Mechanization of Culture

Walther Rathenau's views about technology were far more ambivalent than Sombart's. Rathenau also understood that technology permeated and defined modern culture, but he feared that its mechanistic nature would deaden the vital organic quality of life. A German of many qualities, Rathenau became the model for the modern Renaissance character in Robert Musil's epic novel, *Man without Qualities*. The titles of Rathenau's widely read books, *Die Mechanisierung der Welt* (Mechanization of the world), *Zur Kritik der Zeit*

(Criticism of the times), and *Von Kommenden Dingen* (Of coming things), suggest the thrust of his cultural vision. As an engineer-entrepreneur, head of the Allgemeine Elektrizitäts-Gesellschaft (German General Electric Company), organizer of Germany's wartime industry, and, after World War I, German foreign minister, Rathenau positioned himself at the intersection of modern cultural forces, including electrical technology and science.

Sensitively attuned to the prestige enjoyed in Germany, and especially in Berlin, by those whose cultural interests transcended their professional or commercial roles, Rathenau presented himself to his contemporaries as a cultural critic and person of discriminating taste in literature, art, and architecture. Not without a touch of irony, he suggested that it was strange that he, an innocuous electrical engineer, found acceptance in social circles that embraced great writers and artists.

Early-twentieth-century Berlin shaped Rathenau's concept of a modern culture. Berlin represented the values and nurtured the institutions of a modern technology-based culture. Rathenau experienced and expressed this mix as thoroughly as any other German of his time. So positioned, he contemplated what he called the mechanization and modernization of the world. He had a broad concept of mechanization, associating it with the application of calculation and reason to improving production.Rathenau believed that rapidly increasing population and the related demand for consumer goods were the root causes of mechanization. Germans, he thought, gave so much of their time and energy to increasing production through mechanization that it shaped their culture. Never in the history of the world, he argued, has a single approach influenced such a wide circle of humankind, as has mechanization. Thinking systematically like an engineer, scientist, and financier, he believed that mechanization was bringing about a political, economic, and social revolution that was transforming the world into a global production system.

With a degree in science and in engineering, as well as managerial experience, Rathenau cultivated a complex system-building

13. Social critic Walther Rathenau expressed the complexity of modern Berlin culture as he wrote of its rich mixture of political, social, artistic, and technological components. Edvard Munch, Portrait of Walther Rathenau *(1907). © 2003 The Munch Museum / The Munch-Ellingsen Group / Artists Rights Society (ARS) NY. Photograph © Scala / Art Resource, NY.*

approach. As an electrical industry executive, he coordinated functionally related technologies and organizations to create systems centralizing the production and use of electrical power. His synthesizing mind could comprehend the potential interaction of capital-rich investment banks, government agencies cultivating industry, private companies committed to innovation, unions dedicated to rational deployment of labor, and research laboratories applying science. He envisioned an amalgamation of the world into one continuous system of production, distribution, and consumption. Mechanization and systematization fused in his mind.

Rathenau had a more advanced concept of mechanization than most of his German and American engineering and managerial contemporaries. They continued to think of machines and linear production processes. Their world consisted of linear gears, belts, pulleys, and reciprocating steam engines. Experienced in the realm of electric power systems, Rathenau thought and spoke about electrical circuits and electromagnetic fields as found in generators and high-voltage electrical transmission systems. Interacting components within systems constituted his world rather than mechanical elements related in cause-and-effect sequences. In referring to systems, he used such metaphors as "net" and "web," metaphors employed by those describing globalization and the Internet today. He anticipated transition of modern technology from mechanization to systematization.

Yet the complex Rathenau reacted ambiguously to the mechanized and systematized world that he helped create. He feared that the shaping of human activities and institutions by mechanization would result in a world devoid of human spirit, or soul. He, like Mumford, whom he influenced, believed that the rationality of mechanization would overwhelm the organic character of life. Rationality, order, and control fascinated him, but he also feared that mechanization leaves a moral void in its wake. He was concerned that deterministic systems of production would take on a life of their own and that humans would become components without volition in the systems.

Ultimately, he belittled the creators of mechanistic systems, including himself. The truly great creators, he believed, devotedly, disinterestedly, imaginatively, and intuitively served transcendental causes. He had in mind Christian saints and Jewish cabalists. He hoped that passion and creativity such as theirs would overcome the forces of mechanization.

American Pragmatism

Unlike Rathenau, Spengler, and Mumford, some leading American social critics reacted positively to the technological transformation of culture. Charles Beard, an eminent and prolific historian who taught at Columbia University until 1917, observed in 1930 that no theme, not even religion, stimulated so much discussion among pundits and interest among the public as the meaning and future development of machine civilization. A tempest in a teapot a few years earlier had become a national discourse.

Beard not only influenced other historians, but also the literate public through his enormously successful American history text entitled *The Rise of American Civilization* (1927), coauthored with his wife, Mary. In it they enthusiastically and often eloquently argue that the electrical, internal combustion, and mass-production technology of the second industrial revolution was bringing an era of economic democracy in which all classes would enjoy material abundance. It, in turn, was sustaining political and social amenities conducive to a rounded quality of life.

The Beards' anticipation of the benefits of electrification and automobilization closely resembled Mumford's, but concern about the mechanization of life did not temper the Beards' optimism as it did Mumford's. They, like Mumford, foresaw electrification eliminating the grime and pollution of coal-fired industry concentrated in industrial cities. Electric power transmission would allow industry to spread throughout the countryside. The automobile and radio would bring culture to people so dispersed. The Beards cheered

14. American historian Charles Beard, in contrast to his German peers, celebrated un-reservedly the deep influences of machine technology upon modern culture. Photograph of Beard courtesy of New York State Archives.

on the "machine age" of the late nineteenth and early twentieth centuries. They believed that the "whole scheme" of American life would be altered by machine technology, mass production in particular.

Mass production would deliver an abundance of goods to the lower economic classes. Machines would displace hand and heavy labor, and women would appear in the workforce in increased numbers. Enhanced transportation would distribute standardized mass-produced goods throughout the United States and advertising would spur demand. Yet the Beards acknowledged a downside to mass production. The mechanization of newspaper publishing decreased costs and led to the lowering of intellectual standards in order to reach masses of readers. Standardized goods and services would sustain a standardized way of life that cultural elites disparaged. Standardization also led many Americans to lament the influx of immigrants who did not fit into the presumed American Nordic mold.

On the other hand, the Beards in their survey of America's technology-based culture found Americans generous in their charitable undertakings, which the Beards confidently predicted would bring the benefits of mass production to all classes. Millionaires grown rich through industrialization also endowed universities and museums to raise the cultural horizons of the masses. The Beards believed that machine-age America provided material sustenance for the body and the mind in greater abundance and variety than any nation throughout history.

Americans' enthusiastic dedication to material abundance through mass production resulted in the cultivation of sciences that could be applied to industry while so-called pure science languished. A similar effect could be noted in the universities where industrial management and engineering schools attracted upwardly mobile students.

Progressive experts realizing that America was moving with originality into a machine-based culture looked for new solutions to urban problems, which included poverty, inadequate infrastructure, and irrational growth. City planning became a recognized profession. Abundant steel and electric elevators stimulated the introduction of the skyscraper as a characteristic American form.

Having traced technological and cultural trends in machine-age America, the Beards ventured to anticipate the future rise of American civilization. They dismissed the dire predictions of Spengler-like critics who believed that America and the West had reached their zenith and that the future would bring the mechanization and deadening of the spirit, the demise of creative genius in the fine arts, the coarsening of social relations, and the loss of control of machines that would run rampant.

Instead, the Beards believed that the onward and upward march of democracy coupled with new and wondrous technology would bring the blessings of civilization—material goods, increased utilization of natural resources, health, and security—and that the masses would deal competently with the problems of technology-based civilization. Physical things would be subjected to the remedial response of intellect and spirit.

In an introduction to an edited volume entitled *Toward Civilization* (1930), Beard responded in greater detail to European aesthetes writing in their ivory towers about machine-age America eradicating humanism and religion. He found them scornful of American civilization, which they lambasted as materialistic and devoid of high culture. Besides British authors Hilaire Belloc, G. K. Chesterton, and Aldous Huxley, Beard singled out a German writer, Richard Müller-Freienfels, as representative of foreigners hypercritically condemning America. Yet in responding, Beard acknowledged many of the negative characteristics of his country's civilization.

Müller-Freienfels, author of the *Mysteries of the Soul* (1929), believed that Americans cherished quantification, rationalization, efficiency, and utility while suppressing agreeableness, the emotions, and the irrational. Morality in America, Müller-Freienfels concluded, did not arise from an inward spirit, but out of conformity to mass mores. Americans, ranging from boxers to workers, acted like machines. Machinelike workers outproduce the world, and mechanized boxers become world champions.

In his travels, Müller-Freienfels observed that his American hosts hearing a Beethoven symphony on their radios and phonographs were interested only in the mechanics of reproduction. The arts did not flourish in technology's nation. Pure science, the quest for truth for its own sake, languished. The pragmatic engineer only took interest in reframing the world to produce goods and services.

Favoring intimacy with machines, Americans did not have the fruc-
tifying close associations with nature and long-standing cultural
traditions. Cultural roots, on the other hand, shielded Europeans
from modern machine idolatry.

Hypercritical Europeans found simplistic quantification to be a
hallmark of the mechanized Americans, who believed that bigger is
better. They measured achievement numerically in the workplace,
professional life, and sports. American opinion makers in the press
and on the pulpit and podium spoke and wrote mesmerized by the
spell of the quantitative. They could not penetrate qualitatively the
profundity of life.

Foreign critics saw Americans behaving like machines. They
had become interchangeable in appearance, attitudes, morals, and
mores. Clean-shaven American men and doll-faced, heavily made-
up American girls seemed to have emerged from a Ford assembly
line. In Europe "bourgeois" pejoratively characterized a philistine,
but in America to behave as a middle-class, standardized, good
citizen was virtuous. Good citizens mistakenly believed that they
shaped the political future of the nation, while in fact money and the
press controlled it.

Beard argued that foreign critics distorted the American character. Yet not all of their accusations, he acknowledged, were capri-
cious or irrelevant. He, like they, lamented the physical desecration
of the environment by a production-driven population. He found
huge sections of inner-city New York City, Chicago, and Pittsburgh
shabby and their waterfronts dilapidated. He deplored the desecration of the landscape by billboards and gas station shacks along the
highways. And he lamented the endless row houses built near ugly
factories.

In *The Rise of American Civilization*, Beard had acknowledged that
U.S. literature mirrored the problems of the machine age. Upton
Sinclair exposed the shocking conditions in Chicago's mechanized
stockyards. Theodore Dreiser and Sinclair Lewis pilloried the self-
satisfied middle class satiated with material goods and devoid of
interest in higher culture, while Edith Wharton, James Branch Ca-

bell, and Ellen Glasgow told of the decline of the established classes whose wealth was based on agriculture, shipping, and textiles, and the rise of captains of machine-age industry. Insecure in its new role of social arbiter, new wealth-seeking culture looked to Europe for art, literature, music, and architecture despite the riches and originality of American talent.

Having temporized, Beard then quickly took the offensive. He celebrated the Christian work ethic and idealized American engineers. He reminded his readers that painters had represented the Virgin spinning by the cradle of Jesus, and Joseph sawing wood. Because the Roman Catholic Church canonized many saints for bringing security, comfort, and prosperity to their communities, Beard ventured that the papacy would have named Edison a saint had he made his inventions in medieval times. (Beard forgot that the papacy continued to canonize.) Without virtual saints like Edison, Americans would never have known the material riches of the machine age.

Beard celebrated the machine, inventors, and Christian virtues, but reserved his lavish praise for engineers. He commended them for forcing nature to conform to human will. He likened them to the all-knowing and powerful Creator giving shape to inanimate, chaotic matter. Engineers, he insisted, are capable of even more heroic and highly imaginative, socially benign enterprises—if set free to fulfill their own of values.

He did not make clear, however, from whose values they are to be set free. Corporate America's? He did not venture to say explicitly what he surely knew. Unlike many doctors, lawyers, professors, architects, and other professionals, engineers were mostly employees of large corporations, such as General Electric and General Motors. Designing automobiles and power plants, engineers took directions from managers who responded to shareholders and the market, as they do today. Yet Beard anticipated engineers becoming an autonomous, self-determining fifth estate.

Beard asked prominent engineers to contribute essays in *Toward Civilization*. Elmer Sperry, Lee de Forest, Lillian Gilbreth, and

Michael Pupin, as authors, reinforced his idealized view of engineering values. They were committed to rationality, the scientific method, and large-scale planning that would bring a good life for millions. Free to pursue their ends, American engineers would use science and invention to nurture justice and toleration and free humans from hunger, poverty, and pain. Just as medieval saints cultivated the highest good, the masters of the machine will nurture the best elements of their civilization. The church fathers promised heaven only in the hereafter, however.

Both Beard and Mumford revealed their political naïveté in assuming that scientists and engineers could wrest the direction of public affairs from politicians and industrialists. Neither Beard nor Mumford suggested how the necessary revolution would occur. Beard erred even further in assuming that engineers would nurture the cultural foundations upon which the fulfillment of the human spirit depends.

Other leading American public intellectuals and scholars shared Beard's positive interpretation of machine-age America. Some concluded that America in the twentieth century embodied scientific and technological values that would determine the future of all civilized peoples. America would lead the way to an idealized future. Arthur Holly Compton, who won the Nobel Prize for physics in 1927 and who taught at the University of Chicago, expressed this sanguine view in a 1936 essay entitled "Oxford and Chicago: A Contrast."

Humanistic dons at Oxford with whom Compton had recently dined left him with both a favorable impression of men steeped in literary tradition and with a bemused tolerance for their otherworldliness. They, he decided, would not flourish in Chicago and certainly not in his university. He pitied them for being needlessly fearful of a world increasingly shaped by science and technology, one shaped by his university and Chicago.

Chicago had developed, he observed, free of the tradition-laden European influences that reached to America's northeastern seaboard. Youthfully enthusiastic, Chicago used technology to create a physical environment suited to modern men and women. The University of Chicago, unlike Oxford University, had three of its four divisions dedicated to the sciences: physical, biological, and social. The faculty and students, according to Compton, were discovering the enriching values of a mechanized world. Compton acknowledged that the modern world could ill afford to lose the leaven supplied by Oxford, but the world needed Chicago to adapt traditional values into the yeasty mix of a mechanized society. Chicago nourished a scientific method that probed and defined a plan for a desirable human evolution.

Matthew Josephson, an influential small magazine editor and biographer, also welcomed the arrival of a machine age. In 1922 Josephson declared in his essay "Made in America" that the machine not only brought American economic ascendancy, but also a flourishing art and literature. American intellectuals, artists, and writers ought not to condemn the machine for "flattening" and "crushing" them: they should venerate the machine as a magnificent slave for all Americans, a hardy, thoroughly modern new race. It is not surprising that Josephson later wrote a highly readable biography of Thomas Edison that portrayed him as an American hero embodying the essence of the nation's inventive genius. In contrast, he wrote another book about the capitalists of the era, the title of which was *The Robber Barons*.

Beard, Compton, and Josephson, unlike a post–World War II generation of intellectuals, found technology defining American characteristics that they believed would set an example for, and arouse the envy of, older European nations. America was forging the way into the future. Just as Americans living on the frontier had, in the opinion of historian Frederick Jackson Turner, acquired the defining characteristics of nineteenth-century America, inventors, engineers, and scientists inhabiting machine-age America were acquiring the characteristics of a new America—technology's nation. In the future, however, a relatively simple machine-age civilization would give way to one characterized by systems, controls, and information, the subject of the following chapter.

CHAPTER FOUR

Technology as Systems, Controls, and Information

Large-scale complexity characterizes the post–World War II era of technological systems and distinguishes it from the preceding machine era. Because of this complexity, the control, or management, of technological systems becomes a major problem for engineers and other expert professionals. Finding that control is closely linked to the communication of information, they are among those who helped bring about the information revolution that took root as the twentieth century drew to a close. In the previous chapters, we have considered technology as a means of transforming a wilderness into a human-built world and as a machine for the production of goods. We will now explore its nature as systems, controls, and information and discover that managing it is a major societal challenge.

The Systems Era

The popular press portrayed Henry Ford as the creator of a massive production machine made up of large automobile factories at Highland Park and River Rouge, but Ford saw the core concept embodied in his factories as "system, system, and more system." [1]

---

1. David A. Hounshell, *From the American System to Mass Production, 1800–1932: The Development of Manufacturing Technology in the United States* (Baltimore: Johns Hopkins University Press, 1984), p. 229.

Frederick W. Taylor, progenitor of modern scientific management had a fixation about systems, too. He went so far as to argue that in the past man came first, but that in the future the system must be first. Even before World War II, systematization was becoming a widespread technological activity.

After the war, perceptive observers saw that systems, not machines, were structuring the industrial world. They realized that the United States had left the age of mechanization and entered the era of systematization. Because the systems concept is more complex and difficult to grasp than the concept of mechanization, not only public intellectuals, historians, and philosophers interpreted technological change for a general public, but social scientists, engineers, and scientists did so as well.

Russell Ackoff—a University of Pennsylvania professor, management consultant, and provocative thinker in the field of operations research and systems dynamics—became an insightful interpreter of the systems era. In 1979 he observed that the United States had moved from the machine age into the systems age with all of the entailed changes. The machine age with its mechanical devices could be understood, he explained, through rational analysis associated with problem solving as taught in engineering schools and schools of management. Because the number of variables was severely constrained, problems could be reduced to quantitative dimensions. Problem solvers assumed that they fully understood a machine or a machinelike bureaucratic organization.

Because many complex systems, on the other hand, cannot be fully comprehended, engineers and managers need to be satisfied with a partial, ambiguous understanding. Technological and organizational systems are often so complex, so large, and so heterogeneous that interdisciplinary interactive groups sharing perspectives and information are needed to create and control them. Figuratively taking machines apart to understand them was, according to Ackoff, the key to machine-age thinking. Synthesis is the essence of systems-age thinking.

Todd La Porte, a political scientist at the University of California, Berkeley, expressed similar views about organizations and management in the age of systems. He emphasized, however, systems control. Having spent extended periods in the field observing the management of large systems, including electric power systems and naval aircraft-carrier task forces, La Porte suggested that in order to control these systems, managerial authority had to be distributed among numerous individuals at various levels. A hierarchical management structure feeding information up to a few executives who pass orders down is not, according to La Porte, sufficiently informed and flexible to respond to the complexity of present-day organic systems.

He also found that large technological systems have subsystems and components whose original purpose has been forgotten and whose present function is no longer clear. Engineers and managers, for instance, often do not recall the circumstances under which many electric power system components were introduced and the purposes for which they were originally intended. Mapping urban water and waste systems is often impossible. These systems are so complex and so poorly comprehended that their managers are unable to weed out obsolete components. La Porte's insights help explain why Ackoff referred to systems management as mess management.

Weapons Systems

Many post–World War II weapons systems have the messy, complex character described by La Porte and Ackoff. The SAGE (Semiautomatic Ground Environment) air defense system developed during the 1950s at the Lincoln Laboratory of the Massachusetts Institute of Technology is an example. It involved advanced-design digital computers, a widespread radar subsystem, a telephone network for communication, and a variety of aircraft and missiles controlled from the ground to intercept incoming enemy aircraft. A seminal learning experience for its managers and engineers, SAGE (and subsequent command and control systems) stimulated a systems discourse and approach that diffused rapidly.

The development of missile systems during the 1950s also stimulated technological and managerial changes that spread into the civil sector. The management techniques used by the U.S. Air Force and its corporate partners in developing intercontinental ballistic missiles (ICBMs) initiated a systems engineering approach stressing scheduling and coordination. It proved especially effective for projects involving thousands of contractors and subcontractors. In the 1950s, the U.S. Navy used a systems approach involving computers and software to create the Fleet Ballistic Missile (FBM) System.

The navy's Fleet Ballistic Missile Weapon System integrated a number of semiautonomous constituent components, including a Polaris intermediate-range (2,500-mile) ballistic missile (IRBM), a specially adapted atomic submarine, a missile launcher, and navigation/fire-control capability. To preside over the development of this complex array and avoid existing bureaucracy, the navy in 1955 established the Special Projects Office (SPO), which functioned as a collective systems builder and systems engineer. Staffed by the cream of navy personnel and civilians, the SPO eventually coordinated the activities of about three thousand aerospace, electronic, and instrument manufacturers, as well as university research laboratories that developed components.

In order to coordinate the countless research and development tasks, the SPO required the numerous contractors to report their progress regularly. Finding the incoming information overwhelming, the SPO, with the help of civil consultants, introduced in 1957 a mainframe-computer-based program named Program Evaluation and Review Technique (PERT). In essence, PERT graphically portrayed a flow plan, or dynamic network diagram, on a computer printout. The diagram revealed the interactive development and flow of numerous design, research, and development projects. With PERT, the SPO could observe and respond to progress and delay in each project as the FBM moved to completion. In the opinion of many who used it, PERT showed how computers and appropriate software could control complex systems, a connection that has proliferated and spread throughout the military and civil management world.

Harvey Sapolsky, an MIT political scientist, however, contends that PERT was more a public relations gimmick than a real management tool. According to Sapolsky, the navy invited members of Congress and others controlling purse strings to observe PERT computers in operation and see complex PERT charts, hoping that the inscrutable aura surrounding PERT would forestall informed criticism and justify expenditures.

The Military-Industrial-University Complex

Sapolsky is but one of many academics who criticized the military's increasing presence on the research, development, and management scene. In his farewell address to the nation in January 1961, President Dwight Eisenhower also raised doubts about the military's role and referred to a "military-industrial complex." Besides warning that a large arms industry could have undue economic, political, and spiritual influence in the councils of state, he cautioned against federal military expenditures influencing the research policies in American universities through allocations and contracts. For this reason, I add "university" to the name of the complex.

Numerous academics criticized the complex and tried to keep classified research off-campus. Seymour Melman, a Columbia University professor, in two influential books—*Pentagon Capitalism: The Political Economy of War* (1970) and *The Permanent War Economy: American Capitalism in Decline* (1985)—argued that a state economy presided over by the Department of Defense was displacing and corrupting the traditional market economy in scale and scope.

Tens of thousands of defense contracts supported both industrial corporations and universities. Between 1960 and 1967, the Department of Defense, according to Melman, awarded the Lockheed Corporation, an aerospace giant, more than $10 billion in contracts and General Dynamics $8.824 billion. In 1968 the Pentagon awarded the Massachusetts Institute of Technology $119 million; Johns Hopkins University received $57 million; the University of California, $17 million; Columbia University, $9 million;

Stanford University, $6 million; the University of Michigan, $9 million; and the University of Illinois, $8 million. Military-industrial-university expenditures, according to Melman, were drying up the government's appropriations for physical infrastructure, health, and welfare. As a result, the nation's economy, especially its once-dynamic free market, was in decline. Melman feared that the United States was becoming a garrison state.

Paul Forman, a curator and researcher at the Smithsonian Institution in Washington, rejected the military's claim that it was disinterestedly supporting basic physics research. On the contrary, he decided that the military supported basic research of a kind that gave some promise of military application. Forman contended that American physics during the period from 1940 to 1960 underwent a qualitative change in its purposes and character—an enlistment and integration into the nation's pursuit of military security. In support of his thesis, he cited physicist Merle Tuve, head of the Carnegie Institution of Washington, who said in spring 1959 that government expenditures for basic research in science contributed little to the really basic core of scholarly achievement. Forman concluded that the scientific talent flows where national priorities place incentives of money, prestige, and excitement.

Spread of Systems Approach

Despite the attacks upon the military-industrial-university complex, it has had a lasting influence upon the character of American technology and its management. Tens of thousands of engineers, scientists, and managers who took part in building weapons systems in the 1950s acquired from this learning experience a systems approach to projects. They and other professionals influenced by them increasingly conceptualized the world around them in terms of systems. Formerly, they might have seen an airplane in isolation; now they saw it as part of a system involving airfields, air controllers, fuel depots, and maintenance facilities. Formerly they might have conceived of a highway in isolation; now they placed it in a network of facilities, including automobiles, service stations, and traffic controls.

To manage the creation of large weapons systems, engineers, scientists, and managers developed a family of systems-based managerial techniques known variously as systems engineering, operations research, and systems analysis. Systems engineering usually referred to the management of the design and development of systems, operations research involved the use of quantitative techniques to analyze deployed weapons systems, and systems analysis compared, contrasted, and evaluated proposed projects. In practice, usage varies from these definitional norms.

Influential organizations and a number of universities cultivated a systems approach. The RAND Corporation in Santa Monica, California, a nonprofit think tank founded in 1948, used air force funding to assemble a staff of experienced mathematicians, natural scientists, and social scientists who impressively developed an interdisciplinary approach to systems analysis. RAND proved especially effective in advising the military about proposed weapons projects. Acting as a collective systems engineer for the first Intercontinental Ballistic Missile Project, the Ramo-Wooldridge Corporation, founded in 1953, developed the systems engineering approach essentially as practiced today. MIT's Lincoln Laboratory, the MITRE Corporation, the Systems Development Corporation, the Air Corps Systems Command, and the U.S. Navy Special Projects Office also cultivated systems approaches in the 1950s.

Academics in leading research universities codified and rationalized the practical field experience of these and other organizations. University courses about a systems approach trained a host of professionals in operations research and systems engineering. Advocates of the approach during the 1960s contended that it made possible the control of vast and complex physical and social structures not only of the military, but also of the industrialized civil world. They held that a systems approach offered a rational response to growing complexity.

Urban Systems: Promise and Failure

In the 1960s, alleviating deteriorating urban infrastructures and housing became a priority for the administration of President Lyndon Johnson. It turned to systems-approach experts who believed they could respond to urban problems as they had to military ones. In 1968 Vice President Hubert Humphrey eloquently expressed the administration's enthusiasm for the systems approach when he predicted that techniques used to manage the creation of large weapons systems and space exploration projects could solve pressing urban problems.

Infrastructure systems have long shaped cities, and cities have come to depend upon them. In late-nineteenth-century America, railroad, telegraph, and telephone networks stimulated the growth of cities. Centralized water supply and waste disposal, as well as urban transit networks, added to infrastructure density. By 1900 the networked city had a downside that has become more serious over time. Since infrastructures are interdependent, the failure of one can cascade throughout the networked complex. For example, a loss of electric power will, in some cities, suspend urban transit systems, automobile traffic controls, elevators, and, in some instances, water-supply pumps. The breakdown of telephone service can interrupt many communication networks dependent upon it, including air-traffic control and the stock exchange. Cascading failures can paralyze a city.

Called upon to respond to urban problems by the Johnson administration, a number of aerospace corporations and consulting engineers familiar with the systems approach floundered as they tried to move the systems approach from the military to the urban sector. In their efforts to introduce new or improved infrastructure, the systems experts found that local governments often would not cooperate when a planned system had to spread over several political jurisdictions. Having customarily given priority to technical and economic factors, they had difficulty in accepting and adjusting to political ones. Coping with the city of Philadelphia proved far more difficult than coping with the Department of Defense.

The systems approach also fell into disrepute because managers and engineers designing and deploying large systems, especially urban transportation systems, did not sufficiently take into account the traditional and delicate social fabric of urban communities. Critics cite Robert Moses, New York State and New York City park commissioner and city construction coordinator, as representative of the insensitive system builder. Yet supporters saw him as brilliantly presiding over the construction of parks, parkways, expressways, strategic bridges, public buildings, playgrounds, public beaches, and public housing in the New York City region from the 1930s to the 1960s. Moses set an example that influenced the destiny of countless twentieth-century American cities. Driven by a bold vision, Moses used public works to structure the life of an urban region.

A Yale Phi Beta Kappa and an honors man at Oxford, the physically imposing Moses began his public service as an idealist and ended it, according to biographer Robert Caro, as a ruthless user of power to fulfill his egotistical vision. Moses masterly framed legislation and established organizational structures, especially public authorities that gave him the freedom to function as an autocrat independently of elected officials and their traditional bureaucracy. Posing as a man of reason acting above politics, Moses functioned as a masterful power broker who maneuvered mayors and other public officials from an awesome power base funded by bond issues and highway and bridge tolls.

Other Moses critics focus upon his rude and high-handed taking of houses, destruction of neighborhoods, arbitrary seizure of rights-of-way, and violation of green environments. Moses believed that overconcern about nontechnical matters could paralyze engineering; in other words, his vision took precedent over the interests of ordinary citizens standing in his way. Moses argued that in laying out urban highways, he removed ghettos, but critics contended that he violated living communities by forcing hundreds of thousands of

15. First celebrated for his New York public works projects, Robert Moses later fell out of favor because his projects autocratically violated the physical integrity of neighborhood communities. Photograph of Moses courtesy of the Nassau County Museum, Long Island Studies Institute.

inhabitants from their homes, many into banal massive high-rise housing. He predicted that the high-rises would provide clean, safe, and efficient living spaces, but inhabitants often found the housing cold and forbidding, as well as isolating them from neighbors.

Yet Lewis Mumford, an insightful and frequently negative critic of his beloved Manhattan's architecture and landscaping, found much good to say about Moses in the 1930s. Writing regularly for the *New Yorker Magazine* in a series of "Sky Line" columns, Mumford declared that all around Manhattan the energetic and astute Moses was creating the framework of a new and improved city. Mumford really liked Moses's playgrounds, marine parks, and landscaped highways, especially those that connected to Moses-designed Jones Beach. Mumford found that the landscaping of Jones Beach left him in a state of ecstatic admiration. The beach architecture, however, left Mumford far less than enthusiastic.

Systems Discredited: Vietnam War and the Counterculture

Popular books and articles by scientists and social scientists heightened the reaction against large technological systems. Writing for the *New York Review of Books* with its large influence among academics and liberal intellectuals, John McDermott drew a memorable analogy between civil systems of production and military systems of destruction that the United States deployed in the Vietnam War. Rachel Carson's *Silent Spring* (1962) calls attention to the loss of natural sounds, smells, and vistas as human-built systems of production with their toxic substances displace nature. Biologist Barry Commoner, in *Science and Survival* (1966) and *The Closing Circle* (1971), writes of an environmental crisis caused by industrial pollution of the environment.

In the title of his book *Small Is Beautiful* (1973), economist E. F. Schumacher found an expression that epitomized the reaction, especially among young adults, against large-scale systems of production. The popularity of the *Whole Earth Catalog* (1968) demonstrated the enthusiasm for small-scale appropriate technology. It identified the hand tools of a benign technology useful for shaping a new environment. Many practitioners of the *Whole Earth Catalog*'s philosophy were trying, like the Amish people in Lancaster County, Pennsylvania, to reduce their dependence on large manufacturing and utility systems.

In his book *Soft Energy Paths: Toward a Durable Peace* (1979), physicist Amory B. Lovins persuasively argues that small power plants using renewable energy sources, such as wind, sun, and water, could more efficiently supply energy than regional power plants with their lengthy transmission and distribution lines. Influential interest groups, however, heavily invested with capital and with engineering and managerial skills in existing systems resisted new technologies that would destroy established ones. Innovative environment-sustaining systems lacked the capital and political influence needed to establish themselves in the face of this momentum.

Lewis Mumford railed repeatedly against the military-industrial complex, which he labeled a "megamachine." Megasystems and their elite experts, according to him, deadened the humanistic aspects of life and took society to the brink of catastrophe. In *The Pentagon of Power* (1970), he describes the massive, centrally controlled military-industrial complex as dependent on a priestly, or scientific, elite, monopolizing knowledge in order to ensure its power, glory, and material well-being.

In *Men, Machines, and Modern Times* (1966), Elting Morison, an MIT history professor, eloquently expresses with a biblical cadence his views about large systems. He holds that they have acquired an intricacy, mass, scale, and rate of change making it extremely difficult for individuals to cope. He sees the irony of industrial society mastering the natural environment, but in so doing making a second creation so complex that it defies control. We are no longer, like Job, struggling with God, but struggling, he laments, in a network of our own godlike powers that we have not mastered.

Perry Miller, a Harvard professor of American literature, in an article aptly titled "The Responsibility of Mind in a Civilization of Machines" (1979), ventures that humans face a tragic dilemma as they find themselves in a universe of their own manufacture with which they are unable to deal. He recalls that Gloucester in Shakespeare's *King Lear* could blame the gods for his predicament, but whom, asks Miller, can we blame for the physical corruption of Gary, Indiana, but ourselves. Numerous other academics and professional writers similarly considered omnipresent technological systems out of control.

The Failure of Controls

While the anti-technology values of post–World War II intellectuals influenced a limited segment of the public, well-publicized technological catastrophes heightened anxieties about the functioning of the human-built world. Large coastal oil spills and the urban smog alerts of the 1960s stimulated a rising tide of public sentiment that brought passage of federal environmental laws modifying industrial practices. The nuclear-reactor catastrophe at Three Mile Island in 1979 heightened public concern and stimulated regulations and controls that dampened the spread of nuclear power. The *Challenger* shuttle tragedy of 1986 temporarily derailed the National Aeronautics and Space Administration's bold program of space exploration.

Technological catastrophes raised serious doubts about the capability of engineers and scientists to control technology, a capacity that they claimed to have had during the machine age. Public anxiety heightened during the cold war as engineers, scientists, and managers presided over massive technological projects culminating in the production of thousands of nuclear-warhead-tipped intercontinental ballistic missiles. Furthermore, the memory of the awful destructiveness of the atom bombs dropped on Hiroshima and Nagasaki created in the public mind a specter of technology running amok. Engineers were displaying creative powers beyond the imagination of Faust and Mephistopheles.

There were disturbing failures in combat operations centers designed to detect enemy and to control friendly intercontinental missiles and aircraft. Constructed in 1961, the U.S. Air Force's underground combat operation center inside Cheyenne Mountain, Colorado, experienced alarming software failures. For eight tense minutes in 1979, Cheyenne mistook a test scenario for an actual missile attack, a mistake that could have triggered a nuclear holocaust. In 1980 a computer-chip failure mistakenly alerted the Strategic Air Command against attack. False alerts continued, especially following the installation of new computers.

Sociologist Charles Perrow in *Normal Accidents: Living with High-Risk Technologies* (1984) has called such failures in complex systems "normal accidents." These are most likely, he argues, in tightly coupled systems in which various components interact quickly over rigid connections. An intercontinental ballistic missile, for instance, with propulsion tightly coupled to its guidance system is prone to "normal accidents." In the case of the *Challenger* spacecraft tragedy, a cascading series of failures involved interacting physical components and humans in management agencies. The near-catastrophe at the Three Mile Island nuclear plant resulted from interacting operator and hardware failures. Likewise, the Chernobyl disaster in the Ukraine can be classified as a "normal accident," as well. And unintended electric-supply-system blackouts have followed upon unanticipated cascading interactions involving operator and hardware failures.

Controls and Information

Aware of the failure of controls and the frequency of "normal accidents," engineers and scientists sought to improve control theory and practice. As early as 1948, Norbert Wiener, an MIT mathematical prodigy, explored the close linkage between controls, information, and communication. In his 1948 book *Cybernetics; or, Control and Communication in the Animal and the Machine,* he declares that the

16. MIT faculty member Norbert Wiener understood the interconnection of control, communication, and information and helped pave the way for the information revolution. Photograph of Wiener courtesy of the MIT Museum.

seventeenth and early eighteenth centuries were the age of clocks; the later eighteenth and the nineteenth centuries, the age of the steam engines; and the twentieth century, the age of communication and control.

Wiener conceived his influential cybernetic theory when designing and analyzing gunfire-control devices during World War II. He and his contemporaries designed electromechanical and electronic systems that communicated information electrically to control devices at a distance. They then defined communication as the transmission of information that controls and orders.

Feedback was a central concept in Wiener's control theory. Along with his close friend Arturo Rosenblueth of the Harvard Medical School and Julian Bigelow, an MIT engineer who built mechanical models to demonstrate Wiener's concepts, Wiener defined feedback in a seminal essay entitled "Behavior, Purpose, and Teleology" (1943). Negative feedback, they explain, is the use of controlling signals to modify the output, or behavior, of a machine or organism so that it will reach its goal. Comparison of a machine's or organism's interim state with the path to its eventual goal generates the controlling, or error, signals.

Often feedback signals must be dampened to prevent the machine or organism from oscillating excessively about the path to its goal. Rosenblueth, a neurobiologist, used as an example of undamped oscillations a patient with cerebella disease waveringly and unsuccessfully attempting to raise a glass of water to her mouth. Rosenblueth, Wiener, and Bigelow compared predictive and nonpredictive behavior in feedback systems. An amoeba seeking a moving goal does not extrapolate the path of its goal; a cat, on the other hand, does extrapolate the future position of a running mouse and moves toward that future intersection. Predictive behavior is divided into orders of complexity: the cat predicts the path of the mouse; a person throwing a stone at a moving target makes a second-order prediction by foreseeing the paths of the target and the stone.

Predictive behavior requires at least two coordinates—a temporal and a spatial. Prediction is raised to a third, fourth, and so on order, if the machine or animate organism has additional sensors that discriminate a number of spatial axes. A bloodhound, for instance, does not evince predictive behavior, for it follows only smell. Because humans employ a number of sensors, their behavior can be of a higher order of prediction.

Several years before the publication of *Cybernetics*, Wiener's ideas had already greatly impressed a small interdisciplinary group of scientists, engineers, and social scientists who joined him in several seminars. This interdisciplinary group of "cyberneticians," as described by Steve Heims in *The Cybernetics Group* (1991), worked its way through cybernetic, or control, concepts and took a leading role in disseminating them by an ingenious use of analogy. They believed that cybernetic concepts had general applicability for researchers in a number of fields, including the human sciences. The cyberneticians' discourse about control and feedback and their imaginative analogies anticipated the information revolution. They probed the relationship between control, communication, and information.

Members of the interdisciplinary group included, besides Wiener, Rosenblueth, and Bigelow, John von Neumann, the Hungarian mathematician who helped introduce game theory and later designed a seminal scientific computer; Claude Shannon of Bell Laboratories, who with Wiener pioneered in the development of information theory; Warren McCulloch, a neurobiologist at the University of Illinois Medical School; Walter Pitts Jr., McCulloch's protégé; and Rafael Lorente de Nó of the Harvard Medical School, who brought a background in neurobiology to the interdisciplinary metaphor-generating discussions.

Other notables in attendance included anthropologist Margaret Mead and her husband, Gregory Bateson, a psychological anthropologist. Mead recalls that the interdisciplinary discussions about feedback controls made a strong impression—precise enough to be used in problem solving, but abstract enough to cross disciplinary boundaries. In May 1942 in New York at the seminar's first gathering, Rosenblueth gave such an engrossing and stimulating presentation on feedback controls that Mead did not notice she had broken a tooth.

Rosenblueth's presentation drew on conversations with Wiener and Bigelow about analogies between machines and organisms, especially their common characteristic of purpose. Traditionally, scientists are reluctant to explain actions in terms of purpose, for this assumes knowledge of actions in the future and placing the effect before the cause. Rosenblueth related purpose to negative feedback, or circular causality and control, another approach scientists avoided because the associated math is so difficult. For Wiener,

Rosenblueth, and Bigelow, teleology is purpose controlled by feedback, as distinct from the common association of teleology with final
causes.

A Flood of Control, Communication, and Feedback Metaphors

Because they realized that analogic metaphors provide connec-
tive bridges, the seminar group used control, communication, and
feedback metaphors as well as computer ones to find similarities
among their various disciplines. They generated a flood of metaphors. Wiener continued to stress analogies between machines and
humans. Von Neumann suggested the analogous behavior of electrons in vacuum tubes with neurons in organisms. And Lorente de
Nó saw similarities between the firing of an impulse from an individual neuron and the digital binary processes of computing machines. Their metaphors and talk anticipated the discourses of the
information age.

Others seminar members imaginatively found analogies between machines and organisms. McCulloch in association with the
younger Pitts conceived of analogies relating machines and the
functional organization of the cerebral cortex. In 1943 they described a model of neural nets that soon proved a useful metaphor
for describing digital computers. Their model involved idealized
neurons capable of generating an excitatory or an inhibitory impulse. Separated from other neurons by synaptic gaps, the electrical
controlling impulses do not fire across a gap unless the voltage of
nerve fibers exceeds the threshold voltage of the neuron. When this
occurs, a neuron will then fire electric impulses in a chainlike reaction to other neurons with which it is in contact through nerve fibers
across synaptic gaps. The process suggests the digital switches of an
electronic computer.

Known for his study of Balinese culture and language and his appreciation of metaphors and analogies, Bateson found the cyber-

netics approach richly thought provoking for social scientists. Years later he recalled cybernetic ideas as more profound and dramatic than the concepts associated with the double helix model of Francis Crick and James Watson. He also believed that cybernetics provides an organizing principle for the social sciences as powerful as Darwinian biology. On another occasion, he wrote that the Treaty of Versailles and the discovery of cybernetics were the two most important historical events in his life. He took as his special responsibility the transfer, by means of analogy, of cybernetics concepts from mathematics and engineering to the social sciences, a transfer leading to an emphasis by scientists and social scientists upon systems and information.

Molecular and Developmental Biology

Because some molecular biologists, especially those with physics backgrounds, left feedback out of their explanations for organic growth and reproduction, Bateson considered their approach reductionist. He found them wrongly assuming that a linear flow of control information moved from DNA genes to the human's protein manufacturing "factory." Like Bateson, Evelyn Fox Keller, a historian of science, found many molecular biologists and geneticists taking a reductionist linear approach and avoiding the tough intellectual demands required when analyzing the simultaneous interactions within complex biological systems.

In contrast to reductionist molecular biologists, developmental biologists introduced the circularity of systems, control, and information to explain growth. They conceived of information sources distributed throughout the human organic network, or system. They took their metaphors from cybernetic systems, such as computer networks, instead of the simpler mechanical machine metaphor used by many molecular biologists. In "The Cybernetics of Development" (1957), C. H. Waddington discusses the feedback needed to ensure the stability of human developmental pathways.

He refers to the self-directing systems from which Wiener drew his cybernetic theories. Michael Apter, a psychologist with computer experience, also conceived of developing organisms as exceedingly complex, feedback-laden systems. These and like-minded scientists in essence argued that genes alone do not explain embryogenesis; a self-organizing and self-steering system should be the explanatory model. In so arguing in the 1970s, they were making potentially serious inroads into the central dogma of unidirectional gene action expounded by Crick in the 1950s.

Molecular biologists later resorted to information metaphors. They used terms such as "inscription," "transcription," "translation," "expression," and "transformation." They referred to bits of genetic text and to the transmission, accumulation, and storage of information. Using a biblical metaphor, historian Lily Kay predicted that the master of the DNA molecule will be the one who reads the book of genetic scripture. Yet even resorting to information metaphors did not free molecular biologists from accusations that their approach was reductionist.

Origins of the Information Revolution

Information theory and metaphors not only pervaded biology, but also increasingly infiltrated scientific, engineering, and managerial discourse about communication and control. This discourse in the 1950s and '60s signaled a forthcoming information revolution, which was soon brought about by a confluence of conceptual, technological, scientific, and organizational developments. Like earlier industrial revolutions, the information revolution has multifaceted origins involving unanticipated confluences.

In *The Control Revolution: Technological and Economic Origins of the Information Society* (1986), sociologist James R. Beniger associates an incipient information revolution with the interaction of controls and information, thus recalling Wiener. Crises of control brought by rapid industrialization, the spread of larger and more complex technological, managerial, and business systems, and an increase in "normal accidents" stimulated a quest for control devices that were information dependent.

Other scientists and social scientists also observed the rise of an information-dependent industrial society, especially in the United States. In the 1960s, historian 'of science Derek J. de Solla Price called attention to the exponential growth of information in the realm of natural sciences. Anticipating the use of computers as information processors, in 1945 Vannevar Bush, a renowned MIT academic engineer and science policy maker, published his plans for the "memex," a memory and retrieval device designed to help academics, lawyers, and business executives awash in information to organize and retrieve it. As wartime head of the Office of Scientific Research and Development, Bush observed the wartime federal government increasing the value of printing and reproducing equipment from $650,000 to $50 million within a year.

Economist Fritz Machlup pointed out that as early as 1958 the information sector of the U.S. economy accounted for 29 percent of the gross national product and 31 percent of the labor force. In the 1960s and '70s, books and articles appeared with titles such as "Production and Distribution of Knowledge," "Age of Information," and "Information Revolution." In the 1980s, the Smithsonian National Museum of American History organized a major exhibition featuring computers called *The Information Age*. Information was becoming a driver of far-flung and deep social and cultural changes, much as energy had been during earlier industrial revolutions. Contemporaries were observing and defining the information revolution.

The Revolution's Technical Core

Like the second industrial revolution, the information revolution involves the spread of new pervasive technology and its interaction with existing technological systems. The introduction of electric power, a new form of energy, generated cascading effects as it displaced steam engines in numerous energy-dependent systems including transportation and production. Similarly, numerous technological systems depend on information, especially for communication and control. Introducing a new means of transmitting information also has cascading effects as digital information displaces older forms of information in various systems. Information and electric energy have similar effects and generate comparable developments because both are means of transmission and distribution, in one case, energy and, in the other, information.

The origins of the information revolution have yet to be thoroughly explored by historians, yet even now it appears that the interactive development of computers, semiconductors, and software nurtured the spread of digital information, which in turn led into the revolution. Just as early inventors and developers of electric dynamos, motors, and lamps did not anticipate that these would interact to cause changes so varied and widespread, the early inventors of semiconductors, computers, and software forged ahead without foreseeing that they were involved in a nascent information revolution.

In 1947 Walter Brattain, John Bardeen, and William Shockley at Bell Laboratories patented a small semiconductor transistor to displace the omnipresent large vacuum tube in military and civil communication and control devices. Saying that he wanted to make a million dollars and see his name in the *Wall Street Journal*, not only in the *Physical Review*, Shockley left Bell Labs in 1955 to start a company to develop, manufacture, and market transistors. By establishing a research-and-development start-up company in Stanford University's research park in Palo Alto, California, he initiated a trend the consequences of which he could not have possibly foreseen.

Nor did Jack Kilby of Texas Instruments and Robert Noyce, who in 1959 independently invented an integrated circuit, realize that this would contribute to the information revolution. The device allowed resistors and capacitors to be combined with transistors in

17. Inventor of an integrated circuit, Jack Kilby displayed the creativity that represents the present-day interaction of technology and science. For his innovation, he received the Nobel Prize in 2000. Photograph of Kilby © The Nobel Foundation.

an electronic circuit on a single silicon wafer. Another major break-through came in 1971 with the invention of a microprocessor by Marcian (Ted) Hoff, an engineer at the Intel Corporation. More complex than the integrated circuit, the microprocessor incorporated hundreds of thousands of circuit components dedicated to the logic of calculating or control, thus becoming a computer on a chip.

The information revolution began to take shape as integrated circuits and microprocessors interacted with the development of computers and software. Microprocessors allowed designers in the 1970s to introduce computers smaller than mainframes, first minicomputers, still large by comparison with today's desktop computers, and then personal computers. Initially having limited use by scientists as giant calculators, computers soon served as control devices for weapons and industrial systems and spread into the civil realm as information processors. In the 1970s, production of transistors, integrated circuits, microprocessors, and computers skyrocketed in California's Silicon Valley centered upon Stanford University. Research and development start-up companies proliferated; their innovations fueled the information revolution.

Personal computing took a giant step in the 1970s when engineers and scientists at the Palo Alto Research Center of the Xerox Corporation designed and built the Alto computer. It featured icons, a mouse, and a pull-down menu designed by Douglas Engelbart, as well as a laser printer. For complex reasons, Xerox's commercial design of this innovative computer was not successful. In Silicon Valley in 1977, Steve Jobs and Steve Wozniak introduced the Apple II computer and later a Macintosh with Alto features, a floppy drive, and an elegant and simple architecture.

Initially mistakenly considered peripheral to computer, or hardware, development, software soon became a major component in the evolving information revolution. After IBM introduced a personal computer with spreadsheet software and word processing, *Time* magazine named the computer "Machine of the Year" in 1983. The IBM machine used William Gates's MS-DOS operating system, and his Microsoft Corporation soon became the world's leading software producer.

Designers of minicomputers and personal computers did not intend them to function interconnected, but the Advanced Research Projects Agency (ARPA) of the U.S. Defense Department funded their interconnection in a system named the ARPANET. After 1971 it became the core of interconnected computer networks called the Internet. Use of the Internet expanded dramatically after Tim

Berners-Lee, a scientist at CERN, the European particle physics laboratory, in 1991 made available the prototype for what has become known as the World Wide Web. The usefulness of the Web increased greatly when Marc Andreesen, a student at the University of Illinois, and Eric Bina in 1982 composed a program for a Web browser that allows users to search the Internet effectively.

As in earlier industrial revolutions, inventor-entrepreneurs such as Brattain, Jobs, and Engelbart, pursuing independent ends, were the immediate causes of the information revolution. Like their predecessors, these inventor-entrepreneurs presided over the development of an inventive idea until development culminated in a marketable product. They often established a start-up company to develop, manufacture, and market their invention. Like the independent inventors of the past, they have been responsible for radical innovations through the launch of new systems.

Managerial, Organizational, and Social Changes

Tightly coupled to technological change, managerial practices and organizational forms have evolved during the information revolution. Hierarchy, specialization, standardization, centralization, expertise, and bureaucracy became the hallmarks of management during the second industrial revolution. Flatness, interdisciplinarity, heterogeneity, distributed control, meritocracy, and nimble flexibility characterize information-age management. The organizational culture of Silicon Valley at the epicenter of the revolution has been described as information sharing, collective in learning, informal in communication, fast moving, flexible in adjustments, entrepreneurial, start-up inclined, and thoroughly networked.

Virtual corporations that focus upon management and that outsource manufacturing and other functions to contractors also display the information-revolution organizational style. Because of its rapid and deep interconnectedness, the Internet allows a virtual corporation to function like a systems engineer presiding over a project by scheduling and coordinating subcontractors. Not invested in facilities to manufacture components made by its contractors, a virtual corporation can nimbly shed its contractors and move to another domain.

The information revolution, like the second industrial revolution, locates in particular cities and regions and brings demographic change. New York and Berlin attracted innovators during the second industrial revolution. Silicon Valley, Boston-Cambridge, Austin, North Carolina's Research Triangle, and northern Virginia have become information revolution sites. Futurists have predicted that the home computer and the Internet will also accelerate the movement of population out of the cities and away from transportation nodes into the suburban home. Work can often be done in the isolated home as well as in centrally located offices. During the second industrial revolution, there was a demographic shift away from the coal-rich regions into those with water and electric power. Because computers are not as energy dependent as is heavy industry, computer and software manufacturers tend to locate in campuslike environments, such as Silicon Valley, where young university-trained engineers and managers prefer to live and work.

On the other hand, the information revolution homogenizes places and transforms them into spaces. Places, be they cities or villages, have history, traditions, and local characteristics stemming from physical and human geography. Cities that lose their unique characteristics become spaces. Today cities throughout the world are losing their architectural identity and moving toward becoming spaces because architects are responding to the siren song of globalism, which they believe to be driven by information-revolution technology. They easily resort to a homogeneous global style. Because technology provides unprecedentedly low-cost transportation and communication, architects and planners are not constrained to using local materials and energy sources. Singapore, Kuala Lumpur, and Dallas resemble one another. They have high-rise office buildings, shopping malls selling similar products, fast-food franchises, villas for the wealthy, condominiums for young middle-class professionals, and traffic congestion reaching into the suburbs.

Reactions to the Information Revolution

Humanists, public intellectuals, and artists lamented the constraining influences of large technological systems upon individual freedom and associated them negatively with the Vietnam War and the deterioration of the environment. In contrast, reactions to the information revolution have been decidedly positive. George Gilder has been the most enthusiastic and influential of the celebrants of the computer-driven information age. Before becoming a prophet of the new age, he made a name for himself writing about supply-side economics in his book *Wealth and Poverty* (1981). President Ronald Reagan's speeches reveal that he quoted Gilder more than any other living author. With the publication of *Microcosm: The Quantum Revolution in Economics and Technology* (1989), he changed from a supply-side evangelist to a technoprophet. *Microcosm* caught the attention of the then congressional Speaker of the House Newt Gingrich, whose enthusiasm for computer technology resembled Gilder's.

Before the collapse of many dot-com companies at the turn of the century, Gilder's newsletter in which he singled out highly innovative dot-coms led his devoted and trusting readers to invest in droves in companies about which Gilder spoke glowingly. For a number of years, Gilder held an annual conference called Telecosm at Lake Tahoe, California. His talks about the religious meaning of the emerging information revolution highlighted the meetings. He believes that the revolution is driven by his and other's faith in the future of technology.

Microcosm established a litany for the faithful, a line of prophecy that has shaped later books and articles of those sharing his enthusiasm. Gilder declares that the industrial age has passed and the

United States and Japan have entered upon the quantum age. Isaac Newton's laws describing the macro physical world are now transcended by those of Max Planck and others who explored and explained a microworld of quantum physics consisting of particles and waves. The result, for Gilder, is that machines and material things, which have been the measure of wealth, are being displaced by creative human minds as the measure and source of wealth. He points out that Japan, a barren group of islands with a scarcity of material resources, has become a leading economic power because of the creativity of its entrepreneurs. In the United States, Silicon Valley is similarly a great resource rich in creative minds.

For Gilder, the microchip made mostly of dirt-cheap sand and computer software devoid of material substance are the prime examples of embodied creativity characteristic of the quantum era, or the information revolution. He is fascinated by the fact that the cheap material in the microchip is given great value by the creativity of designers. In celebrating the newness of the new age, he chooses to ignore the fact that for centuries engineers have transformed inert natural materials into economic resources.

The designers of microchips and the other technology of the new age, he delights in saying, are not Ivy League graduates in gray flannels and button-down blue shirts, but outsiders, nerds, science wonks, and upwardly mobile engineers. The acne-faced, ponytailed young people who work seventy hours a week have carried the United States into the information age. Alongside them are the immigrants from Madras, Israel, and Malaya.

The creators of the new age who generate wealth by creativity are becoming not only the masters of the economy, but also of politics and social life. Using computer telecommunications, they are the destroyers of large bureaucratic government and industrial organizations. One entrepreneur sitting at a computer workstation, according to Gilder, exercises more world-transforming power than the captains of heavy industry sitting atop a massive hierarchical structure.

Gilder's appeal to the young hackers, virtual-reality denizens,

Net surfers, multimedia artists, and other information-age enthusi-asts is understandable. As he observes, they are not Ivy League cor-porate aspirants: they are the outsiders who are becoming the in-siders who live, or perish, by their skill in developing new hardware and software. They, like Gilder, dismiss the past as irrelevant and believe that their world is entirely new under the sun. They fer-vently believe that computer-driven technology will change every-thing. Vested interests in the status quo are ignored.

Gilder's messianic and technically informed style engages the new age creators. Manuel Castells's *The Rise of the Network Society* (1996), a denser study, influences the way in which social scientists and humanists understand and react to the information revolution. After acknowledging the hype surrounding the majority of books and articles about the information-technology revolution, he, nev-ertheless, equates its impact with that of the British industrial revo-lution. A professor of sociology at the University of California, Berkeley, Castells argues that both revolutions in a period of several decades thoroughly transformed the material basis of society and changed its culture.

Castells focuses upon the networked global economy emerg-ing in the last quarter of the twentieth century. Interconnected computers throughout the world enable producers and consumers, borrowers and lenders, investors and brokers, to instantaneously ex-change information. The exchanges transcend national bound-aries, so the constraints of national economies give way to an inter-dependent global economy. Because of the primacy of information as the new raw material and creator of wealth, world regions pros-per or decline not so much because of natural resources, but be-cause of the capacity of their managers, engineers, scientists, and workers to harvest knowledge as raw material. The global economy supports an international division of labor that locates regional manufacturing of computer components where knowledge and skill reside. Regions without these assets languish.

Castells's most original ideas deal with what he calls the space of flows. He imagines a global electronic network superimposed upon the world, a network along which digital information consisting of texts, images, and voice flows instantaneously. Interactions are often simultaneous rather than sequential. This network supersedes the railways, highways, and communication linkages of the pre-information age. Global manufacturing, commercial, and financial firms function within this space of flows.

Besides the network, nodes exist in the space of flows. Global cities in which information is generated and managed become network nodes. They are losing their local connections and becoming spaces that resemble one another. Elites presiding over organizations and institutions that control flows on the network choose to live and work in the global cities. Their high-rise business centers, luxury hotels, and airports all tend to resemble one another.

In *The Closed World* (1996), media historian Paul Edwards provides an enlightening perspective on Castells's space of flows. Edwards defines the closed world, or cyberspace, as an artificial space inside a computer or a computer network. In this space, nothing exists except abstract nonphysical information. The action in science-fiction books such as William Gibson's *Neuromancer* and films such as *Blade Runner* and the *Star Wars* trilogy take place in a closed world, such as cities and space stations, which are devoid of animals and plants, in short, devoid of nature. Humans and cyborgs inhabit the spaces. Cyborgs (cybernetic organisms) are integrated humans and machines, especially computers, that usually have the forms of either robots or humans. Dramatic tension in the books and films often involves conflict between cyborgs and humans. A computer network, not unlike Castells's space of flows, often provides a constraining structure within which action takes place. Edwards argues that these fictional closed worlds distill and simplify our anxieties and aspirations in the so-called real world.

Both Castells's space of flows and Edwards's closed world share a disconnection from nature, or what Edwards calls the green world. Humans sometimes try to escape from the closed world into the green one. While rationality prevails in the closed world, natural forces, emotions, community, and even mystical, magic powers prevail in the green world. The space of flows and the closed world, on the other hand, are human-built, human-imagined worlds completely disconnected from physical nature and completely controllable.

Tom Forester's *High-Tech Society* (1987), like so many other books on the subject, breathlessly characterizes the information technology revolution as dramatically transforming society. *High-Tech Society* is representative of the hype to which Castells alludes and that Edwards avoids. Forester, an author or editor of five books on technology and society, approvingly quotes others who describe developments in computer hardware as the most remarkable technology ever confronted by humankind and the digitalization of information as the twentieth century's most fascinating development.

Technological developments in conjunction with deregulation and privatization have resulted, Forester believes, in new companies and new products emerging as never before in history. New telecommunications systems involving computers, digitalization, and fiber optics, Forester is assured, will bring changes as momentous as the railroad and highway systems. Computers in the factory will revolutionize industrial production; in marketing and finance, the changes will be comparable. He ventures that the speed of the high-tech revolution is much faster than that of the industrial revolution. After surveying high-technology developments in the United States and Japan, Forester predicted in 1990 that Japan would become the world's leading economic power by 2000.

Bill Gates, the head of the software giant Microsoft Corporation, touts the bright future of the information revolution. In *The Road Ahead* (1995), he shares his vision of a future for which, he says, he can hardly wait. He confides that he has been doing all he can to bring it closer. Not surprisingly, Gates, as a successful marketer, stresses that the revolution promises a huge market for goods, services, and ideas. It will change what all of us buy and how we invest. It will also determine who our friends will be and how we will spend our time with them. The workplace and education will be transformed beyond recognition. If that were not enough, the information age will create a new human identity. In short, he concludes, almost everything will be different.

In *Being Digital* (1995), MIT professor Nicholas Negroponte, the founding director of the MIT Media Laboratory, is fascinated by the transformation of various kinds of information into a common digital form. He is apparently unaware that electrical power during the second industrial revolution similarly transformed various forms of energy, such as steam, water, and chemical, into a common electron form. He compares bits, the unit of digitalization, with the atoms of the past. More appropriately, he might have compared bits with electrons, but then the digital revolution he is describing would not seem as dramatic and unprecedented.

Negroponte optimistically embraces the information revolution. After acknowledging that jobs may be lost to computerized automation, that individual privacy may be violated, that digital vandalism will occur, and that large sectors of the population will feel disenfranchised, Negroponte moves on to predict that like an irresistible force of nature, digitalization will decentralize, globalize, harmonize, and empower. A new generation of "kids," unconstrained by the need to be in geographic proximity, will collaborate over digital networks to increase global harmony. Like so many technological optimists in the past, Negroponte believes that people who know each other better will like one another better, too.

He concludes that the digitized tomorrow will exceed people's "wildest" predictions—a statement that can be taken several ways. To temper Negroponte's enthusiasm, we should recall Thoreau's gentle reminder that the dappled sunlight falling across the path of the poet provokes joy beyond that which technology can bring. The gentle wind cooling the heated brow fills the poetic mind with profit and happiness equal to that which inventions supply.

Enthusiastic authors arguing that the information revolution would change everything for the better helped alter public attitudes toward technology. Having gone along with humanists and social scientist who damned systems in the 1960s, the public now viewed technology more favorably. The profits being taken from technology stock did nothing to dampen their ardor. Because weapons technology employed during the heavily televised Gulf War of 1991 seemed to be the major reason for a low-casualty U.S. military victory, memories of ineffectual, stalemated technology during the Vietnam War apparently faded into the past. Time will tell how the crash of dot-com stocks and terrorism will affect attitudes toward technology. Dramatic swings in attitudes stimulated by passing events have long characterized history.

CHAPTER FIVE

Technology and Culture

We have seen technology utilized to create a human-built world, observed its deployment as a machine for production, and considered it as the source of systems, controls, and information. Now we shall observe twentieth-century architects and artists employing technology as a means to make architecture and art and using concepts borrowed from technology aesthetically as signs and symbols.

The attitudes of architects and artists toward technology changed during the twentieth century. Initially, machine technology fascinated many of them, especially Germans. When technology became associated in the minds of American artists and architects with oppressive systems and controls, they reacted with hostility, but as the information revolution took hold, their attitude became more positive.

We shall explore the work of Germans and Americans during the machine era, because the second industrial revolution stimulated highly innovative work on the part of both. Afterward, however, our attention will be focused on American architects and artists because they reacted more innovatively and interestingly than Germans to the era of systems and to the information revolution.

Machine-era German Architects

Influential German architects believed that machine technology was creating a new modern culture. They aspired to construct architecture expressing modern values. They believed that the modern metropolis with its railway stations, harbors for mighty ocean liners, factories, and electric power plants, rather than cathedrals and palaces, manifested the character of modern times. The eminent German architect and industrial designer Peter Behrens argued in 1910 that technology should not be an end in itself, but used as a means to establish a culture expressing itself through the language of art.

German architects used machines and modern management methods in construction, which they borrowed from the United States. They considered the United States as a technologically advanced country, but not one in which architecture was transforming technology into a modern culture. Le Corbusier, the eminent French architect, spoke for many German avant-garde architects when he said look to American engineers for ideas about architecture, but not to American architects. Lewis Mumford agreed and asked when, if ever, American architects would develop their own modern style.

In their designs, German avant-garde architects used visual machine metaphors to articulate a modern machine-based culture analogous to the religion-based culture that prevailed in the medieval West. German architects and industrial designers, including Behrens, Hermann Muthesius, Walter Gropius, and Hannes Meyer, saw machine technology as both a prime mover of social change and a source of values shaping design. Their artifacts, factory buildings, and housing settlements manifest a machine aesthetic embodying machine rationality, efficiency, symmetry, and clarity of function. In the construction of their massive housing settlements designed for workers, they used building techniques dependent on machines and depended upon industrial management practices.

Muthesius—a pre–World War I German architectural theorist, critic, and designer—argued that not only buildings, but also products in a machine era ought to be designed for machine production. As a leading member of the German Werkbund, an organization of industrialists, architects, designers, and artists dedicated to improving industrial design, Muthesius exerted considerable influence. The shapes of traditional handicraft, he reasoned, were not suited for efficient machine manufacture. As a homely example of a product designed for machine production, Muthesius referred to contemporary clothing that was simply cut and sewn. Flat surfaces, right angles, cubes, spheres, and other geometric forms, he pointed out, suited modern production better than the undulating and convoluted organic shapes favored by the craftsperson and the *arts nouveaux* designer. Technical and economic efficiency also mandated the stripping of decoration from artifacts and the use of inexpensive human-made materials such as glass, concrete block, and structural steel.

Behrens impressively fulfilled Muthesius's call for machine-shaped design. As artistic and architectural adviser to Berlin's Allgemeine Elektrizitäts-Gesellschaft (AEG), a world-renowned electrical manufacturer, Behrens distinguished himself as a designer of factories, worker housing, electrical appliances, and advertising graphics. He defined for future professionals the essence of an inclusive industrial design.

His AEG turbine factory (Turbinenfabrik), completed in 1910 to house machines for the construction of turbines, remains a world-famous architectural monument. With an exterior suggesting the shape of a turbine, the building led a contemporary critic to exclaim that a logical and unified building becomes a symbol of that which it encloses. Behrens designed other factory buildings for AEG, including a high-voltage equipment factory (1909) and a small-motor factory (1913), both of which have also become industrial monuments. Like the turbine factory, these massive buildings have handsome proportions, clean lines, and ample glass.

After the company designated him its artistic and architectural adviser in 1907, Behrens took this rare opportunity to establish the

*18. Peter Behrens's broad range of activities as architect and industrial designer for Allge-
meine Elektrizitäts-Gesellschaft justifies calling him an artistic director. He has been a role
model for countless European designers. Drawing of Behrens (1911) by Max Liebermann.
© Bildarchiv Preussischer Kulturbesitz, Berlin.*

overarching, modern face of AEG. Products designed by him
reflected the style of modern industrial culture. His cladding for arc
lamps was not only aesthetically attractive, but also suited for mass
manufacture. Following Muthesius's precepts, he used plane surfaces and geometric forms. His arc lamp offers a visual analogy to
the lamp's mode of manufacture and function.

In 1910 Behrens summarized his design philosophy in a seminal
lecture entitled "Art and Technology." He maintained that machine and electrical technology were ushering in a modern indus-

trial civilization and that he and like-minded architects and designers had the responsibility to create appropriate architectural forms
and aesthetically pleasing industrial buildings representative of a
modern culture. In his effort to achieve forms that derive directly
from the machine, Behrens anticipated the work of Ludwig Mies
van der Rohe, Le Corbusier, and Gropius, major twentieth-century
architects who apprenticed in his atelier.

Gropius, Meyer, and the Bauhaus

In 1919 Gropius founded the famed Bauhaus school in Weimar and
moved it in 1925 to Dessau, where it is located today. The school
and Gropius flourished as Germany stabilized economically and as
socialist local governments funded large housing projects designed

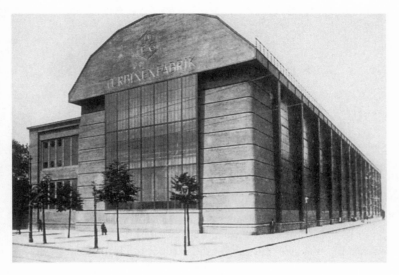

*19. Architects today recognize Peter Behrens's design for a Berlin turbine manufacturing hall
as an industrial architectural masterpiece. Peter Behrens, Turbinenfabrik, courtesy of Lan-
desarchiv Berlin.*

20. Peter Behrens's greatly admired design for an arc light expresses the values of an era when machine technology prevailed. The turbine hall stands in the background. From "Brochure Cover: AEG Flammeco-Lampen," courtesy of Tilmann Buddensieg.

by leading modern architects. The Bauhaus attracted a talented faculty dedicated to improving industrial design of house furnishings and to using advanced industrial techniques, such as those associated with Ford and Taylor, to build houses economically.

In 1924 Gropius called for mass production of a machine-made house, or "*Wohnford*" (house Ford).[1] He wanted factories to produce

1. The expression comes from the architectural historian Siegfried Giedion. Gilbert Herbert, *The Dream of the Factory-Made House: Walter Gropius and Konrad Wachsmann* (Cambridge: MIT Press, 1984), p. 4.

standardized, interchangeable house components that could be assembled rationally into various combinations. In automobile parlance, Gropius advocated different house modules. In an essay, "How Do We Build Cheaper, Better, More Attractive Dwellings," Gropius summarized his approach: the use of mass-production methods; capital-intensive, laborsaving, special-purpose machinery; factorylike assembly-line processes; the division of labor; and flow charts and other management-control techniques. He conceived of the Bauhaus school as a research laboratory for the construction industry.

A derivative of the Bauhaus design became known as the International Style because its proponents believed it transcended national borders and traditional styles. They considered it the essential modern style, one likely to survive into an endless future. The Nazis, however, shut down the Bauhaus school in 1933, believing it to be un-German. Instead, they cultivated and nourished German romanticism and kitsch. After World War II, however, Bauhaus International Style spread rapidly throughout the United States, where many rectangular high-rise buildings simply reduced the once-innovative style to a money-saving formula.

Hannes Meyer, who succeeded Gropius in 1927 as director of the Bauhaus school, revealed in his manifestos and designs the attitude of left-leaning architects in Germany and the Soviet Union toward machine technology and modern architecture. He and they felt confident that machines as tools and symbols would help usher in a modern socialist world. After heading the Bauhaus, Meyer moved to Moscow as a professor of architecture.

Earlier in his 1926 manifesto, "Die neue Welt" (The new world), Meyer conveyed his highly sanguine attitude, one compounded of technological enthusiasm and socialism. Meyer concluded that his generation was witnessing the mechanization of the globe and the domination of thinking humans over amorphous nature. Consequently, values and physical forms were undergoing drastic changes. As evidence of these, he provided a catalog of machines that shaped the everyday world, a listing that became in the mid-1920s a litany of progress for machine enthusiasts. He celebrated Ford automobiles, Fokker and Farman airplanes, Fordson tractors, Burroughs calculating machines, mobile homes, railway sleeping cars, ocean liners, and His Master's Voice phonographs.

Meyer also described a house as a machine for living in and proclaimed that constructing buildings was a technical, not an aesthetic, process. Function dictated design. Fresh air, sunlight, electric lighting, hygienic facilities, garages, and laborsaving kitchens became for him the hallmarks of modern housing. Grain silos, factories, office buildings, and airports absorbed the energies of Meyer and the new architects. New building materials and new techniques heavily dependent on machines facilitated the construction of housing settlements. Provincial styles of building gave way to an international one. For Meyer, the machine and modernity were one.

American Organic Architecture

Paradoxically, leading American architects designing in the world's most advanced technological nation failed to anticipate and develop a style incorporating the values of a machine age. At the close of the nineteenth century, Louis Sullivan, a brilliant Chicago architect who contributed substantially to the rise of the steel-framed high-rise building, spoke admiringly of technological progress, but his designs revealed first and foremost his attachment to the organic. His early reading of natural evolutionists Charles Darwin and Herbert Spencer became a formative experience for Sullivan, who conceived of buildings as organisms shaped by their physical and cultural environment. His influential dictum that form should follow function essentially called for organic forms that evolve from their functions. The availability of structural steel, electric lights, and electric elevators made possible his high-rise buildings, including the Wainwright Building in St. Louis (1890–91), but technology was only one shaping force in an environment, which

21. *Although he flourished when the United States became the world's preeminent technological power, Louis Sullivan designed with organic, not machine, metaphors in mind. He decorated the Wainwright Building with nature symbols. Wainwright Building photographed in 1907 by Emil Boehl. Courtesy of the Missouri Historical Society, St. Louis.*

also included the rising cost of urban real estate and urban commercial expansion.

Sullivan failed to articulate an embracing concept of a modern technology-based culture. The Promethean American transformation of a wilderness continued to shape Sullivan's philosophy of architecture. When he wrote of the modern, he had in mind an era that began in the Renaissance, not the delimited modern era of the second industrial revolution. He used Darwinian and Nietzschean metaphors instead of technological ones to convey his philosophy of architecture and to describe his work. Sullivan adorned his buildings with symbols borrowed from nature. He differed from the International Style in that his major commissions were for commercial buildings rather than industrial ones.

Sullivan's younger contemporary Frank Lloyd Wright has become America's best known architect, but not as a formulator of the modern International Style. In a 1901 essay, "The Art and Craft of the Machine," Wright came tantalizingly close to calling for a new technology-based culture and an architecture expressing machine values. He spoke of mind giving form to the bronze, steel, and plastic of the machine age. He called an electric lamp a machine. In an extended metaphor, he designated Chicago a machine, but he also regretted that its architecture did not reflect the character of modern men and women. Like European avant-garde architects of the era, he lamented the continued eclectic use of construction techniques and forms from the classical, Gothic, and Renaissance past. He disparaged commercial skyscrapers with granite decorations like those once cut by Greek slaves.

Yet he considered technology as essentially tools and machines for construction, not as a source of values to be expressed in design. Reminding architects that they lived in an age of machines and not slaves, he urged them to use machines as tools. He chided the followers of John Ruskin and William Morris, who wanted to revive handicrafts, for shunning the use of modern tools and for using machines to make vitiated copies of handcrafted originals. Lamenting that a vast machine-ridden urban environment suffocated city

22. *Like Louis Sullivan, Frank Lloyd Wright in his early years designed houses that reflected the nature of his regional prairie environment. Wright's Frederick C. Robie House, Chicago, Illinois, view from southwest (1910). Photographer: Judith Bromley. Courtesy of the Frank Lloyd Wright Preservation Trust.*

dwellers, Wright characterized Chicago as a lusty material giant devoid of ideals, shorn in tattered garments, and motivated by greed. His urban vision differed sharply from German architects' appreciative vision of a modern Berlin.

Like Sullivan, young Wright did not design factory buildings. In an article in the *Ladies' Home Journal* entitled "A Home in a Prairie Town," Wright signaled his move toward an anti-mechanistic prairie style. Between 1901 and 1910, he masterfully designed a number of prairie-style homes with their organic metaphors, nature analogies, and contextual respect for natural sites. In the early twentieth century, major architects of technology's nation let their European peers define a technological culture. The natural frontier continued to fire the imagination of American architects.

German Artists and the New Objectivity

In the 1920s, an attitude and an artistic style inspired by mechanization spread in Germany. Labeled the Neue Sachlichkeit (New Objectivity), it reflected the disillusionment following upon the horrors of the lost war and subsequent political and economic chaos. Liberal Germans turned away from idealism and romanticism and opted for social stability and economic prosperity that they believed existed in the down-to-earth, commonsensical, production-oriented United States. They envied Ford's mastery of mass production and Taylor's scientific management techniques. A productive, mechanized culture depended, Germans believed, upon a scientific, pragmatic, value-free, or objective, approach to economic, political, and social problem solving.

The German reaction to the Tiller Girls, an American dance review troupe, suggests the extent to which Germans embraced the mechanization of their culture, believing that it would bring a new objectivity. The precisely choreographed dance routine of the Tiller Girls led intellectuals to refer to them as mechanized humans. Today we would call them cyborgs. Their dance routine reflected the orderly production rhythm of an efficiently run factory and resembled the repetitive motions of workers harnessed to a smoothly flowing assembly line. The Tiller Girls' appeal indicated the triumph of the rational capitalistic values of systems, order, and control over sentimental romanticism, as expressed in a lyrical waltz. Their scantily clad, athletic bodies harmonized well with the modern emphasis upon sport and fresh air.

A contemporary group of German painters adopted the new objectivity. In a 1925 German exhibition entitled *Die Neue Sachlichkeit*, their work displayed their matter-of-fact, or objective, approach. Their style emphasized a hard, metallic sharpness that dispensed with confusing complexity, lacked traces of brush strokes, and was rigidly structured and static. Their compositions, however, were not simply representational engineering or architectural renderings.

The artists reorganized machines and industrial structures, high-lighting elements in order to capture more effectively what they believed to be the essence of the technological artifact.

Grossberg

Among influential Neue Sachlichkeit artists, Carl Grossberg effectively developed the technology theme and established the machine as a major subject for painters. He painstakingly delineated both the systematic, interacting components of machines and the factories in which they were located. Trained in architecture, his paintings suggest, but do not slavishly resemble, architectural drawings. No grime, oil, rust, or workers introduce messy complexity into a world of pristine technology. Grossberg's compositions strongly recall toy-train layouts or engineers' models of industrial complexes.

Industrialists commissioned some of his early works portraying machines and factories. Between 1923 and 1931, however, in a series of dream, or vision, paintings, Grossberg introduced a discordant note by occasionally suggesting surrealistic phenomena. In *Kessel in einer Raffinerie* (Boiler in a refinery) (1933), he ostensibly represents the apparatus of industrial chemistry, but he does not indicate its overall purpose or how the technology works. The looming boiler form and a stepped container beneath it suggest a great icon hung above an altar. This impression recalls Bertolt Brecht's poem "700 Intellectuals Worship an Oil Tank" in which the playwright ironically captures the awe in which some German intellectuals held technology.

In *Traumbild: Dampfkessel mit Fledermaus* (Steamboiler with bat) (1928), a disconnected boiler sits mounted like a museum artifact. One bat hovers above while another sits motionless in a stripped-bare space. According to German folklore, bats are demonic and in dreams portend death, so Grossberg may be symbolizing the end of technical progress. In *Maschinensaal* (Machine room) (1925), the artist shows a Madonna representing the world of religious tradition

23a. Carl Grossberg's Kessel in einer Raffinerie *(Boiler in a refinery) (1933) suggests a factory interior as a place of worship. Courtesy of Eva Grossberg.*

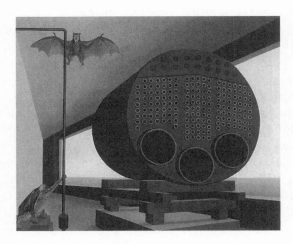

23b. Carl Grossberg's Traumbild: Dampfkessel mit Fledermaus *(Steamboiler with bat) (1928) has symbols that raise doubts about the future of technology. Courtesy of Eva Grossberg.*

23c. Carl Grossberg's Maschinensaal *(Machine room) (1925) contrasts the richness of religion and the sterility of a mechanized world. Courtesy of Eva Grossberg and Von der Heydt-Museum Wuppertal.*

that has given way, because of such inventions as the printing press (pictured in the foreground), to a cold, modern, mechanical world seen through the windows in the painting. Sitting upon the printing press, a monkey represents, as in folklore, the devil. In this and several other of his dream paintings, Grossberg, like Spengler, insinuates that the creation of machines reveals demonic hubris and challenges the role of God the creator.

Machine Art in America: Duchamp

Other European artists also fell under the spell of mechanization, especially American mechanization. Among them were the Dada artists, who painted and exhibited in Zurich, Cologne, Hannover, Paris, and New York City during and immediately after World War

I. Because of its reaction against artistic traditions and bourgeois values, Dada engaged interestingly with the newness, or modernity, of American technology.

Seeing American technology up close deeply shaped the Dada art of Swiss-born Marcel Duchamp. He left Paris during World War I to settle in Manhattan, where he became the energizing center of a group of artists known as New York Dada. In the United States from 1915 to 1923, Duchamp used machine imagery in one of his most celebrated and least understood works, which he inexplicably and inscrutably titled *The Large Glass* or *The Bride Stripped Bare by Her Bachelors, Even*. It stimulates varied interpretations, despite the extensive, but opaque, explanatory notes left by the artist. Yet it is a cornerstone in the extended spread of machine art of the time and merits detailed consideration.

The upper half of *The Large Glass* represents the "Bride." It contains forms analogous to human organs, as well as organlike chemical utensils. X rays of the body, increasingly common at the time, may have suggested to Duchamp an image that descends vertically on the left side of the work representing the Bride stripped bare to her skeleton. An amorphous shape at the top center suggests a cloud of soft flesh, which Duchamp called "Milky Way." In essence, the Bride is shown as fluid and organic in contrast to the geometrical and mechanical Bachelors below.

Duchamp designated the lower part of *The Large Glass* the "Bachelor apparatus." Prominent among the mechanical images are a paddle wheel and a coffee grinder. *The Large Glass* may suggest that males have dispensed with females (the female has become a skeleton) and turned to machines to satisfy erotic needs. Visual sexual metaphors appear frequently in *The Large Glass*. Duchamp and other Dada artists imagined themselves and inventors of machines playing a godlike role in creating "girls without mothers" and making machines in their own image.

In a detailed and imaginative study, *Duchamp in Context* (1998), Linda Dalrymple Henderson shows that Duchamp frequently referred through visual metaphors to contemporary science and

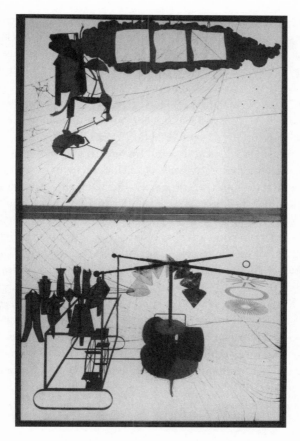

24. Marcel Duchamp's The Bride Stripped Bare by Her Bachelors, Even (The Large Glass) *(1913–15) mixes scientific, mechanistic, and organic images in an impenetrable composition. Oil, varnish, lead fuse wire, lead foil, and mirror silvering on cracked glass panel. Courtesy of the Philadelphia Museum of Art: Bequest of Katharine S. Drier. © 2003 Artists Rights Society (ARS), New York / ADAGP, Paris / Estate of Marcel Duchamp.*

technology. Besides mechanical references, Duchamp introduced concepts of invisible forces and a supersensual world. In his notes he referred to radiation from the Bride influencing the Bachelors below. He described the Bride as existing freely in a fourth-dimensional context while the Bachelors are constrained to the ordinary three. Besides X rays and radioactivity, his visual metaphors refer to non-Euclidean geometries and electromagnetic waves. He associated the "Great Glass" with "Playful Physics."

His notes also reveal interest in electricity, including radio, or wireless telegraphy, as well as incandescent light and electric power generation and transmission. Electromagnetism stimulated his imagination because of its sexually suggestive language, which refers to excitation by sparks and vibration. There are visual references to electrical condensers and wireless telegraph antennae. Duchamp's use of fuse wire as a material to make *The Large Glass* suggests that he is wiring his work for electrical transmission. The Bachelors, or Malic Molds, resemble Crookes, or cathode ray, tubes.

Admiration for Leonardo da Vinci, an artist-engineer, may have inspired Duchamp's wide-ranging and imaginative explorations of technology and science. He also referred to himself as an "artist-engineer" and was familiar with the imaginative drawings of engineering artifacts and natural phenomena in Leonardo's notebooks. Duchamp even declared in 1921 that he was abandoning art and becoming an engineer—neither of which he did.

Sheeler

Duchamp and other European artists seeking wartime refuge in Manhattan often gathered at the apartment of wealthy and generous art patrons Walter and Louise Arensberg. There, European artists mingled with American artists, including Charles Demuth, Charles Sheeler, and Morton Schamberg, who shared an interest in machine civilization and Dada art. Known as precisionists and immaculates, they contributed to the rise of an American style of painting variously called machine art and cubist realism. Their subjects included technological artifacts and urban and industrial scenes, especially New York and its skyscrapers, Pittsburgh with its steel mills, and Detroit with its automobile factories.

Sheeler and Schamberg studied at the Pennsylvania Academy of the Fine Arts in Philadelphia, then one of the world's leading industrial cities. For a time, they shared a studio in Philadelphia, but the Arensberg circle drew them to Manhattan, where they fell under the influence of Duchamp and Francis Picabia, another Dada artist from Europe. Influenced by the latter's machine drawings, Schamberg in 1916 completed eight canvases with machine iconography. The Arensbergs hung one of his paintings prominently in their sitting room.

Schamberg had also studied architecture at the University of Pennsylvania, and a contemporary critic described his drawings as geometrical diagrams resembling a combination of architectural plans and engineering drawings, thus recalling Grossberg. Sewing and bookbinder's stitching machines were among the models for his work. Before he died of influenza in 1918 at the age of thirty-seven, Schamberg was planning many more machine drawings.

Sheeler, too, knew industry and understood machines. Besides attending the Pennsylvania Academy, Sheeler studied at the School of Industrial Art, which was attached to the Pennsylvania (now Philadelphia) Museum. Founded in 1876, the school endeavored to improve the design of machine-manufactured products and to raise the artistic sensibility of workers.

Sheeler looked for the beauty inherent in useful things. Both a photographer and a painter, Sheeler did numerous detailed renderings of factories and machines. Free of grease and grime, his pristine works, like those of Grossberg, call to mind the harmony and detail in an engineer's abstract mechanical drawings. Yet in Sheeler's representations of the machine, there is none of the ambiguity toward technological progress found in Grossberg's dream paintings. Sheeler expresses the unreflective American enthusiasm found in Charles Beard's writings, while Grossberg recalls the Faustian doubts of Spengler.

Sheeler is best remembered for his paintings and photographs of America's preeminent production complex—Henry Ford's River Rouge plant. In 1927 Edsel Ford, Henry's son, commissioned Sheeler to do a series of River Rouge photographs, several of which Sheeler later used as models for paintings. Rather than automobiles, Sheeler photographed the buildings and the machinery inside. Designed by the American architect Albert Kahn, the buildings had clean lines and geometric proportions found in the modern style. Their steel-framed, glass facades expressed the aesthetics of the machine age.

Sheeler found Ford and Kahn's mammoth industrial complex functioning like a single system. He saw individual machines as synchronized components meshing with one another like gears to form a vast system of incredible efficiency. In panoramic views of River Rouge, Sheeler silhouetted railways, cranes, conveyors, storage silos, stacks, and piping against a bleak sky. Only white smoke from stacks suggests active production. His industrial, or technological, landscapes ironically recall the natural landscapes of early-nineteenth-century American painters. He reacted to River Rouge, the human-built sublime, much as his nineteenth-century predecessors did to Niagara Falls and the Rocky Mountains, the natural sublime.

Sheeler used "landscape" in the title of two of his River Rouge paintings. *American Landscape* (1930) portrays the ship slip, ore and limestone bins, storage silos, the cement plant, the iron- and steel-making buildings, as well as rail lines and conveyors. In an oil painting entitled *Classic Landscape* (1931), he focuses on the cement plant and the "High Line" railroad used to transport heavy materials. Only one small human figure stands in these rigid, spare paintings portraying the dynamic flow of materials at River Rouge.

Sheeler reacted like a dedicated engineer to efficient machines. Highly tuned automobile engines brought him pleasure akin to that he derived from hearing Bach because parts in both work together so beautifully. The human mastery of enormous power fascinated Sheeler. For him, the imposing modern industrial landscape stimulated an emotional reaction akin to a religious experience. Like European architects, he equated the factory with the Gothic cathedral;

25. *In Charles Sheeler's painting of the Ford River Rouge Plant, the human-built has become an American landscape. Compare with the wilderness landscape in Thomas Cole's* Tornado *(fig. 6). Sheeler,* American Landscape *(1930), oil on canvas, 24 × 31 inches.* © Museum of Modern Art, New York. Gift of Abby Aldrich Rockefeller, 1934.

both captured the spirit of their age. Machine worship, he believed, was replacing traditional religion.

For decades Sheeler's River Rouge series influenced the manner in which admiring artists portrayed industrial America. The gifted American photographer Margaret Bourke-White, for example, shared Sheeler's passionate response to the machine sublime. For her, industry was an inspiring, vital source of art. A machine's beauty arose from its elegant and functional simplicity.

Bourke-White

Bourke-White's father—an amateur photographer, successful inventor, and engineer—introduced her to machinery at an early age. Born in 1904, she became an independent photographer when technological enthusiasm peaked during the machine age. She was a role model for professional women as she displayed courage, ambition, entrepreneurship, and winning charm. (When under pressure from businessmen because of her casual cost accounting, she responded with an invitation for negotiations over "cocktails.")

Like Sheeler and other artists and architects of her era, Bourke-White found beauty in machinery and industrial processes. Her approach captured the Gothic shapes and awesome power of intricate machinery. She established herself as an industrial photographer after she persuaded the president of Otis Steel in Cleveland, Ohio, that steelmaking, an essential American industry, had a magnificent beauty, which her photography could capture. He gave her free access to the mill, an unheard-of privilege for a woman. In photographing slag-bearing ladles, she rode traveling cranes, persisting despite incredible heat. Otis Steel published her striking smoke-filled photographs, and she won first prize in a Cleveland Museum of Art exhibit in 1928.

Commissions followed from the Ford Motor Company, International Harvester, Republic Steel, and others, but a major breakthrough came when the publisher of *Time* magazine Henry Luce, aware of her photographic homages to technology and business, invited her in 1929 to join the staff of his new, expensive, handsomely published magazine, *Fortune*, which covered technology, business, architecture, and modern design. Her photographs soon documented numerous *Fortune* narrative essays. The magazine then sent her to photograph German industry, and while there she artfully persuaded authorities in the Soviet Union to allow her to photograph the country's mammoth Five-Year Plan projects, including the construction of Dneprostroi, the world's largest hydroelectric dam, and Magnitogorsk, the world's largest steelmaking complex. For the first time, she focused upon workers, portraying them as other than inanimate machine tenders.

Her reputation increased dramatically, when she signed an exclusive contract in 1936 with Luce's new magazine, *Life*, which

26. Margaret Bourke-White's photograph of the hydroelectric generators at Niagara Falls portrays the congealed energy of one of the world's great power sites. Bourke-White, Niagara Falls Power: Hydro-Generators (March of the Dynamos) *(1938), 17⁵⁄₈ × 13¹⁄₈ inches. © Estate of Margaret Bourke-White. Courtesy of George Eastman House.*

became a sensational success. Her picture of Fort Peck Dam in Montana appeared on the cover of the first issue, which also included some candid shots of engineers, workers, and women living at the construction site. Archibald MacLeish, the eminent poet, wrote the captions.

In later phases of her career, she married the writer Erskine Caldwell (who later authored *Tobacco Road*), photographed German air attacks on Moscow, became a war correspondent with the U.S. Air Force, took part in bombing missions, and reported on the Italian front. She fell prey to Parkinson's disease in the 1950s, which ended her professional career, and died in 1971.

Mechanization Takes Command: Giedion

Not American architecture or art, but its design of machine-made home furnishings and appliances, as well as its industrial production techniques, inspired Siegfried Giedion, a Swiss historian of design, to write a seminal work aptly entitled *Mechanization Takes Command* (1948). He concluded that the recent history of mechanization could be best observed in the United States. Yet he was no advocate of mechanization, which had accelerated, he believed, out of human control and without a moral compass. Paraphrasing Thomas Carlyle, who wrote that fine arts had fallen into insanity and walked abroad without keepers, Giedion declared that mechanization did so, as well.

Giedion found that that Americans, following the lead of the British Victorian middle class, deployed mechanization tastelessly in the manufacture of highly ornamented home furnishings. Mass-produced furniture and objet d'art, once finely done in marble and wood, were being cast and shaped in cheap iron, papier-mâché, and rubber. Manufacturers of upholstered overstuffed furniture indiscriminately used cheap, imitative textiles. Earlier, in the hands of master craftsmen working in a strong tradition and responding

27. Siegfried Giedion, second from left, *not only wrote a history of mechanization, but also proselytized on behalf of a modern style of architecture. Courtesy of the Institut für Geschichte und Theorie der Architektur (GTA) ETH Zürich: Nachlass Siegfried Giedion.*

to the cultivated tastes of aristocratic and burgher classes, ornamentation had not run amok. Once machines displaced artisans and produced inexpensive and crude copies of handcrafted work, everyday objects and public monuments became grotesque imitations expressing the acquisitive, envious, undiscerning materialism of a rising middle class.

Contemptuous of tasteless ornamentation, Giedion wanted practical-minded Americans to express their positive democratic and gender values by using machine-made home appliances as laborsaving devices. Like Catharine E. Beecher and her sister Harriet Beecher Stowe, he believed that this would enhance the authority and self-esteem of women. In their influential book entitled *The American Woman's Home* (1869), the author of *Uncle Tom's Cabin* and her sister Catharine declared that the management of the household was a female profession, not simply a chore. As profes-

sional managers, women should use machine technology to mini-
mize their labor and need for servants, the use of which became an
anomaly in a democratic society.

Giedion also favored the views of Christine Frederick, as expressed in a series of articles entitled "The New Housekeeping" in
the *Ladies' Home Journal* (1912). She maintained that the availability
of electrical appliances hastened the mechanization of cleaning,
laundering, ironing, and dishwashing. Recalling that for years she
had made countless wasted motions in doing her household tasks,
she also advocated scientific management in the home, following
Taylorist principles of work found in mechanized factories.

Mechanization, Giedion maintained, affected farming even
more than the household. As in the case of assembly-line production,
changes in agriculture appeared most dramatically in the United
States. Tractors, reapers, binders, and harvesters transformed the
virgin soil of the prairie into commercialized, specialized farmlands
of enormous acreage. He marveled that a railroad traveler in the
evening sees the sunset on the cornfields of Illinois from his Pullman
window and awakes the next morning still surrounded by them.
Giedion quoted California farmers who pointed out that they no
longer raised wheat, but manufactured it.

Giedion feared, however, that the forces of mechanization
would prove more difficult to control than natural ones. Consequently, the twentieth century could become one of mechanized
barbarism, unless the Western world shifted from unbounded, rational mechanization to holism and organicism. The spread of rel-
ativistic and quantum physics, as well as a systems approach to biology, gave, he felt, some evidence of such a shift. He endorsed
Holism and Evolution (1926), written by Jan C. Smuts, prime minister
of the Union of South Africa, which challenges the simplistic, ana-
lytical, rigid, mechanistic approach.

American Industrial Design: Loewy

Between the two world wars, a small group of American indus-trial designers tried to avoid "mechanized barbarism." They de-fined themselves as a profession dedicated to improving the design of mass-produced products. Raymond Loewy, Walter Dorwin Teague, Henry Dreyfuss, and Norman Bel Geddes rank among the pioneers. The life and work of Loewy, especially, offers insights and understanding of the industrial design profession that has been so deeply influenced by modern machines and technological systems. Loewy and his associates' approach differs in critical ways from the design approach that Europeans, especially Behrens and the Bauhaus designers, practiced earlier.

Born in France in 1893, Loewy attended a preparatory school emphasizing engineering and science. He enjoyed the courses and looked forward to an engineering career, but the death of his par-ents and the outbreak of World War I frustrated his plans. During the war, he served as an officer and was cited several times for brav-ery. After the war, he emigrated to the United States and found work as a window dresser for Saks Fifth Avenue and Macy's and as a fashion illustrator for *Vogue* and *Harper's Bazaar*. Wanting to be re-membered as an engineer by training and an artist by profession, he founded a consulting firm specializing in industrial design in the late 1920s. He soon obtained commissions from several manufacturing firms, which brought him in touch with engineering again.

Glenn Porter, an organizer of a Loewy exhibition at the Hagley Museum in Wilmington, Delaware, in 2002, points out that Loewy flourished as the United States became the world's leading prac-titioner of consumer capitalism. Looking for ways to persuade Americans to consume what they mass-produced, manufacturers turned to mass advertising and industrial design, as defined by Loewy, who realized that an America style of design should differ from the European. In America, industrialists produced for the masses; Europeans produced for the upper-middle class. America was an economic as well as a political democracy. While Henry Ford mass-produced a car that his workers could afford, European automobile producers supplied luxury cars for an affluent market. And while utility owners in London and Berlin catered to the well-to-do or to industrial consumers, Samuel Insull, head of Chicago's Commonwealth Edison utility, wanted all classes to use electric lights and home appliances. Ford, Insull, and other American industrialists democratized desire.

Loewy developed a style he called "contemporary American" that blended the modern with the traditional. Wanting people to feel at ease, as well as modern and adventuresome, Loewy designed convenient-to-use products and large and comfortable spaces. Loewy believed that design should make the manufacturer profit, the user happy, and, if possible, not offend the aesthete.[2] His business clients appreciated his measuring the success of a product's design by its sales. Even though his associates took part in much of the design work, Loewy often publicly took full credit. He knew, like Edison before him, that a celebrated name associated with a company or laboratory becomes a brand attracting support.

Frigidaire, Coca-Cola, Shell, Greyhound, and others awarded his company important contracts, but it and he became most famous for streamlining Pennsylvania Railroad locomotives and Studebaker automobiles. Like so many designers, artists, and architects during the machine age, Loewy and his associates assumed that a functional design would be a beautiful one. They considered streamlining, or aerodynamic design, the functional look of the machine age. Loewy even streamlined a pencil sharpener.

The 1930s to the '50s were Loewy's greatest years. In 1934 he became a consultant for the Pennsylvania, the world's most renowned railroad. Before Loewy, the Pennsylvania Railroad had a reputation as a reliable, profitable, mechanically impressive, but stodgy company run by engineers, even though it employed renown architects

2. Glenn Porter, *Raymond Loewy: Designs for a Consumer Culture* (Wilmington, Del.: Hagley Museum and Library, 2002), pp. 6, 11, 34.

28. *Because it allowed him to combine the aesthetic with the technological, Raymond Loewy especially enjoyed designing the streamlined envelope for the Pennsylvania Railroad's S-1. Loewy with the S-1 at the New York World's Fair (1939), courtesy of the Hagley Museum and Library, Wilmington, Delaware.*

such as Frank Furness to design train stations and McKim, Mead and White to do Penn Station in Manhattan, which a subsequent generation thoughtlessly demolished. After Loewy became its design consultant, the Pennsylvania and the Loewy office collaborated on images and objects that symbolized the modern era.

After the Pennsylvania asked Loewy to streamline the GG-1, the railroad's new diesel electric locomotive, he set out to prove to the railroad's crusty engineers that he was no longhaired artist. He did not intend to "pretty up" a 6,000-horsepower locomotive, but to prove himself a realistic and practical designer. His handsomely streamlined GG-1 became universally admired. Loewy went on to design Pennsylvania's widely acclaimed S-1 locomotive. After watching a million-pound S-1 pass at 120 miles per hour, Loewy felt overwhelmed by a feeling of the machine's power. He took pride in what he had helped create and contributed to his adopted country, musing upon the long way that he had come since his start in fashion advertising.

In designing automobiles, Loewy sought elegance, finesse, and emotional appeal. A lover of fast, sleek automobiles, he and his firm styled a cleanly streamlined, swift, and low-slung 1953 Studebaker, at a time when other American cars were large, overdecorated, and fin-tailed. In 1962 Loewy unveiled the Studebaker Avanti, his most daring auto design. Car aficionados loved it and the press praised it, but the Studebaker Company failed to move the car into mass production, turning out only thirteen hundred in an entire year. Loewy quipped that no one could expose an exciting body to the public and then deliver nothing for ten months—except possibly Brigitte Bardot. Plagued by declining sales, Studebaker gave up the automobile business in 1967.

After the 1950s, Loewy's fortunes and reputation declined too. Looking back in his eighties over the history of industrial design, Loewy regretted that he and his respected fellow professionals had failed to rid mass production of ugly American things. They had not prevented corrupt designers from applying cosmetics to shoddy, profit-making products that ushered in a neon and plastic civilization. Loewy lamented that profit-seeking industrialists continued to reject his first-class designs, which offered limited returns, and instead chose inferior ones that promised greater profits. Consumer capitalism was overwhelming quality industrial design.

Artists against Systems, Order, and Control

Celebrated by designers and artists during the machine era, technology during World War II and the cold war became associated in the opinion of many artists with weapons and industrial pollution. They reacted strongly against technology's systematic order and control by stressing chance and disorder. Highly respected artists, including Robert Motherwell, Willem de Kooning, Barnett Newman, Jackson Pollock, and Robert Rauschenberg, wanted their spontaneous art to be an antidote for the dehumanizing impact of highly automated, controlling technology. De Kooning contrasted the individual artist's deep engagement in the production of art with the depersonalization and alienation of the worker in industrial production. Newman, an anarchist candidate for New York mayor, opposed the orderly system of corporate capitalism, celebrating instead spontaneity.

An influential art critic, Clement Greenberg, whose essays nurtured abstract expressionism, observed that these painters withdrew to a domain they believed to be uncontaminated by technology, science, and industry. Meyer Schapiro, another major art critic, similarly pointed out that artists were communicating with a new visual language distinct from the commonplace one used by technological media such as television. They used a visual language requiring an interaction between the artist and the viewer instead of a language that assumes passive viewers receiving controlling messages.

Schapiro argued that abstract expressionists demonstrated a commitment to craftsmanship that celebrates freedom of expression. Paintings and sculptures, he wrote, are the last handmade personal objects within our culture. Artists could find a personal identification in their work, as few others are able to do in an industrial society fostering abstract mass-produced artifacts devoid of a personal touch.

Pollock's seemingly random drips of paint and de Kooning's freely moving, big brush strokes reveal their emotional involvement and their embrace of accident and disorder. Avoiding naturalistic representation, they expressed through signs, symbols, color, line, and space their deeply felt counterreactions to the values of what they believed to be a repressive, systematized industrial society.

Cage

John Cage, a composer, painter, and poet, also provided antibodies against the spread of rigid order and control. His musical compositions, highly experimental and innovative, often called for free improvisation by performers within a loose Cagean frame. He utilized a variety of unconventional sounds, such as cowbells, closing doors, and the hum of electrical generators, as well as silence. His composition entitled $4'33''$ has a performer seated at a piano, playing nothing for four minutes and thirty-three seconds. Cage's poetry presents hexameters incorporating words randomly, and his paintings stem from his chance encounters with found objects.

Cage did not attempt, like many artists of an older generation, to bring order out of chaos. He did not seek to improve the world, as do engineers, but affirmed life as it is. His father, incidentally, was an inventor and engineer. Cage's abhorrence of control extended across a broad spectrum of categories, including grammar. He compared syntax to oppressive government; both have to be obeyed. He chose linguistic strategies that overwhelmed intentionality without annihilating it. Anarchy, he observed, dissolves the bonds of regulation while still permitting individual responsibility.[3]

3. N. Katherine Hayles, "Chance Operations: Cagean Paradox and Contemporary Science," in *John Cage: Composed in America*, ed. Marjorie Perloff and Charles Junkerman (Chicago: University of Chicago Press, 1994), p. 238.

29. *Willem de Kooning in his broad brush strokes and seemingly random composition ex-*
presses the spontaneity designed to counter the order and system of a technologically structured
world. De Kooning, Composition *(1955), oil, enamel, and charcoal on canvas, 79⅛ ×*
69⅛ inches (201 × 175.6 cm), Solomon R. Guggenheim Museum, New York (55.1419).
© 2003 The Willem de Kooning Foundation /Artists Rights Society (ARS), New York. Pho-
tograph by Robert E. Mates, © The Solomon R. Guggenheim Foundation, New York.

30. Not given to order and control, John Cage preferred to let the pieces fall where they may in his art and music. Photograph of John Cage lying on a cobblestone street by Guido Harari. Courtesy of the John Cage Trust.

Katherine Hayles, a professor of literature,[4] finds it passing strange that Cage, who sought assiduously to avoid exercising control and being controlled, frequently referred to his works as "chance operations." She sees a contradiction between the randomness and lack of control that chance suggests and the purposefulness of operations performed to achieve a goal. While chance escapes our designs, operations put our designs into effect.

Hayles suggests that this seeming contradiction inherent in the oxymoronic phrase "chance operations" may be partially resolved by consideration of Cage's vision of intersecting world lines. Cage borrowed the idea of world lines from science-fiction writer Stanislaw Lem, who in his book *A Perfect Vacuum* defines chance as an intersection of world lines. These intersections are for Lem—and Cage—unplanned and, thus, uncontrolled conjunctions, or

4. Hayles has a doctoral degree in chemistry from the California Institute of Technology.

chance. On the other hand, world lines are causal chains associated with human operations. "Chance operations" are, therefore, a combination of conjunction and human cause.

At least superficially familiar with Claude Shannon's information theory, Cage attempted to maximize the content of his poetry by using words randomly. Shannon pointed out that the content of messages could be compressed by substituting code for customary and repetitive combinations. Following Shannon, Cage believed that randomness in his compositions avoided the compression that regularity and repetition allowed and maximized content. He balanced random elements with conventionally ordered prose so that the baffled reader would not give up the search for meaning.

Cage used *I Ching*, the Confucian book of changes, to set limits within which chance events occur. Thousands of years earlier, Chinese scholars utilized the *I Ching* with its hexameters to suggest the nature of the universe, assuming that it was ordered by numerology. For Cage, throwing *I Ching* sticks was like throwing dice, drawing cards, or some other process resulting in a random display. The random display for him became a set of instructions, parameters, or guidelines within which he composed music and within which musicians performed.

An outstanding example of Cage's chance operations music is *Europeras*, performed at the Frankfurt, Germany, opera house in 1987. From a reservoir of traditional operatic arias, stage settings, and costumes, he randomly chose a mix resulting in the performers simultaneously singing arias from various operas, wearing costumes from operas different from the ones they were singing, and performing against a stage setting assembled randomly from a number of operas.

Some of his musical compositions depend on chance conjunction as, for instance, when each among a group of performers plays or sings according to his or her own timeline. There is no need for imposing order and system, Cage suggested, because patterns, repetition, and variations spontaneously appear, like timeline conjunctions. He opaquely remarked that time's one-way directionality affirms the predominance of chance over operation. Cage's time is fractal time, exemplified by the way in which different locations in glass crystallize, or age, at various rates. He used, as well, computer programs to randomly choose the guidelines governing a musical composition.

Cage also staged events, anticipating the "happenings" of the 1960s. In 1952 at the experimental Black Mountain College in North Carolina, he surrounded an audience with sound from a phonograph hand-operated by artist Robert Rauschenberg along with dances by Merce Cunningham, poetry reading, live music, and slide projections. A few years later Allan Kaprow, an American painter, stimulated by the abstract expressionist painters' emphasis on spontaneity and by Cage's chance and indeterminacy, staged happenings encompassing a heterogeneous mix of visual projections, random sounds, and various kinds of performers. Happenings occurred often in the studios of American artists and in hippie communities.

Flourishing in the 1940s, bebop musicians also cultivated the aesthetic of spontaneity. Influenced by an African American musical idiom, Lester Young, Charlie Parker, Miles Davis, and Dizzy Gillespie favored a call-and-response, or conversational, music. Young explained that music emerged because each player had a spontaneous awareness of voices within and without. Young, pointing to his head, told his fellow musicians that things were going on up there, but "man" some of you guys "are all belly." [5]

Best played in small ensembles and featuring improvisation, bebop contrasts starkly with the controlled and orderly style of both symphonic music and big-band swing. A symphony orchestra performs in a large hall before an audience compartmentalized hierarchically according to the price of seating, while bebop musicians play in intimate surroundings in close contact with an audience.

5. Young quoted in Daniel Belgrad, *Spontaneity: Improvisation and the Arts in Postwar America* (Chicago: University of Chicago Press, 1998), p. 179.

While bebop is free and exploratory, a conductor directs a symphony orchestra following a score and exercising control. Big bands of the 1950s also played in an orderly and well-drilled fashion. Carefully rehearsed band members followed the written scores of arrangers. Only featured band members did carefully orchestrated solos. In contrast, the mood and rhythm of bebop passed from soloing musician to soloing musician.

Architects React against Order and Control

Some leading American architects in the 1950s and '60s also reacted against systems, order, and control. Denise Scott Brown, an influential urban planner and architect, championed freely evolving urban places. She urged her peers to learn from the vernacular architecture of Los Angeles, Las Vegas, and urban sprawl, where ordinary people of various tastes materially expressed their aspirations without controlling direction from conventional professional architects and planners. Architects, she believed, could learn from freely evolving vernacular and express its spontaneity in their designs. She, her architect husband Robert Venturi, and their associate Steven Izenour publicized their views in *Learning from Las Vegas* (1972), a book that stimulated a stormy reaction from the architectural establishment, mostly negative.

Scott Brown wanted architects to break free from the overwhelming influence of wealthy, bourgeois clients who commissioned expensive homes, sat on the building committees of large corporations, and on the boards of Ivy League schools. She anticipated that her approach would spread as young architects and planners began to learn from commercial and popular culture and to listen to social scientists who analyzed the attitudes and aspirations of the varied users of buildings and spaces. So informed and inclined, architects would be retained, Scott Brown believed, by citizen groups to help them design projects.

Art, Architecture, and the Information Revolution

In the 1990s, the spread of computers and the information revolution tended to dampen the negative reactions of architects and artists to technology as a source of constraining systems, order, and control. They began to see the computer as a remarkable creative tool enabling architects to introduce highly complex architectural forms and artists to resort to innovative art techniques, including computer graphics, movie animation, and virtual reality.

Like early computers and software, innovative computer-mediated architecture and art originated in university environments. In 1963 Ivan Sutherland, who studied and did research at MIT, introduced the Sketchpad, a seminal, interactive graphics system for manipulating two- and three-dimensional objects. Not only engineers, but also artists and architects found the Sketchpad extremely useful. After moving to the University of Utah as a professor, Sutherland and his professorial colleague Dale Evans created a computer science department that trained graduate students in interactive graphics. Its alumni introduced graphics software, including RenderMan, which was used in making *Toy Story*, the first all computer-animated film, as well as the dinosaurs in *Jurassic Park* and the cyborg in *Terminator 2*. Hundreds of computer-literate artists and model builders were soon creating animated virtual-reality special effects as seen in such popular films as *Titanic* and *Saving Private Ryan*.

The U.S. military has become a major incubator for computer graphics and simulations. A positive-feedback relationship exists between the designers of military battlefield and air-war simulations and the designers of commercial video games, special effects for cinema, and computer-animated movies. The military-entertainment complex took a quantum leap in 1999 when the U.S. Army awarded a $45 million, five-year contract to establish an Institute for Creative Technologies at the University of Southern California (USC). The institute represents a merger of the military, Holly-

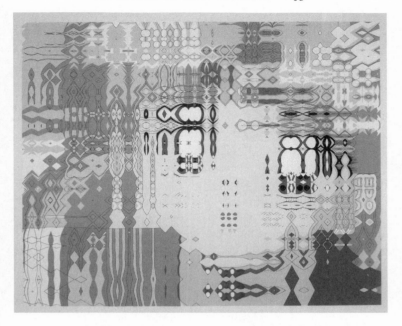

31. For screen print Rhapsody Spray 1 *(2000), Carl Fudge manipulated and digitalized an image of a character from Japanese anime called* Sailor Moon. *Courtesy of Ronald Feldman Fine Arts, New York. Photograph: Hermann Feldhaus.*

wood, and academia intended to integrate their common interest in simulations and virtual reality for civil and military purposes.

Artists are drawing upon the software of computer-using graphic designers and animators. This cooperation may influence art in ways comparable to the manner in which linear perspective influenced painting and design during the early European Renaissance. The Whitney Museum in New York City and the San Francisco Museum of Modern Art both had digital art shows in 2001. The artworks at the Whitney often recall abstract and minimalist styles that are familiar to the experienced museum visitor, styles done in the past with conventional paint, acrylic, pastels, and watercolors. The show's originality stems mostly from the artists' use of computers and software as tools to create the art. Some of the artists programmed their own software, but usually they used available software for manipulating scanned images and for generating three-dimensional solid models and animated images. As their familiarity with the digital tools increases, they will find new forms that are especially suited for the new tools.

Architects now design buildings using graphics software originally intended for automobile styling and movie animation. Wes Jones, a Los Angeles architect, calls the results "blob" architecture. "Blob" refers to designing spaces using the computer's capacity to manipulate three-dimensional forms as if made of soft putty. A change in one dimension reverberates throughout the entire model. The noted architect Frank Gehry, who designed the highly praised art museum in Bilbao, Spain, takes a slightly different approach. He models his buildings initially in a soft material and then turns the model over to computer experts, who use CAD (computer-aided drafting) and CAM (computer-aided modeling) software to develop computer models and plans for the engineers and construction artisans to follow. Gehry's office also employs the CATIA software developed by the aerospace industry to transform intricately sculptured models into engineering drawings. Empowered by technology, Gehry has created a new architectural language of cloudlike forms.

Venturi's and Scott Brown's architecture also broke away from the orthogonal, geometrical buildings of the International Style. Venturi's Vanna Venturi/Hughes House (1964) in the Chestnut Hill section of Philadelphia initiated a widespread and sustained reaction against the order and system of the technology-driven International Style. Contradicting the renowned International Style architect Mies van der Rohe, whose buildings expressed the subtle, clean, and orderly lines of the engineering world, and who said "less was more," Venturi asserted that "less was a bore." Instead of order and control, Vanna Venturi/Hughes House expresses complexity and contradiction.

32. Frank Gehry's design for the Stata Center at the Massachusetts Institute of Technology shows how an architect with computer software can abandon conventional geometry and explore complex and seemingly free forms.

33. In the early 1960s, Robert Venturi designed a house for his mother that cast off the stereo-types of the International Style of architecture and ventured into a realm of aesthetically grounded complexity and contradiction. Robert Venturi, Vanna Venturi / Hughes House, rear elevation. Photograph by George Pohl. Courtesy of VSBA.

The Demise of Technology-based Culture

Despite the recent tendency of digital technology to soften the negative reactions of artists and architects to technology, it is unlikely that the world will ever witness the reappearance of an enthusiastically sustained, technology-based culture of the kind that flourished in the United States and Germany between the world wars.

Instead, artists, architects, and the concerned public have begun to doubt that a completely human-built world can respond to human needs and aspirations. The long-standing fear expressed by Mumford and others that the mechanistic will deaden the organic has risen to the fore again, but in a new guise. A closed computer world is now the mechanistic and the natural environment is the organic. Perhaps the resolution of this tension will come in the form of an ecotechnological world, which we will explore in the next chapter.

CHAPTER SIX

Creating an Ecotechnological Environment

Today people in the industrialized nations, especially the United States, do not grasp the large range of possibilities for creative action that technology offers. We are satisfied to see it used mostly for consumer goods and military weaponry, not realizing that we are unconsciously and unthinkingly also using technology to create a human-built physical environment. We do not take responsibility for the aesthetic characteristics and quality of life in this human-built world. In short, we do not understand the range of technology and our responsibilities for it.

Technologically empowered, we have reason to doubt our values and competence as creators of the human-built and as stewards of the remaining natural world. Slums in inner cities, ugly strip malls in the suburbs, hastily and cheaply built housing, polluted air and water, the loss of ecologically nurturing regions, and the likely threat of global warming give evidence of our failure to take responsibility for creating and maintaining aesthetically pleasing and ecologically sustainable environments.

We also fail to take responsibility for the creation of an eco-technological environment, which consists of intersecting and overlapping natural and human-built environments. More ecologically sensitive and technologically empowered today, we should ask engineers, architects, and environmental scientists to negotiate with one another as they design and construct the ecotechnological environment.

Our forebears could blame God or the gods for floods, droughts, vicious storms, and other natural disasters, but who are we to blame but ourselves for the degradation of the physical environment and its forces that constrain and demean the human spirit? We usually deny responsibility by pointing accusatory fingers at developers, architects, engineers, urban planners, and politicians. They, however, are influenced by our moral values, for we commission, fund, and elect them to fulfill our goals. There are ways that we could participate in the decision making that shapes public and private places, but most of us fail to use the opportunity. To participate, we would need to inform ourselves about the processes involved in designing, developing, and constructing the ecotechnological world, but we are not technologically literate.

As previous chapters have shown, technology in the past has been used to create a human-built world, to make machines for production, to create large systems, especially ones dedicated to information, and to provide tools and subject matter for artists and architects. Perhaps in the twenty-first century, engineers, environmental scientists, and the public will use technology to respond to global ecotechnological problems. If so, then technology will be deployed in complex, even messy, ways because the problems will be laden with political, economic, social, and aesthetic values. Such messy problems require a multifaceted, complex approach. A reductionist approach that is limited to technology will not respond adequately to the problem. Such a reductionist approach is rightly labeled a "technological fix," which has negative connotations. Unfortunately, Americans with a powerful array of technology tend toward technological fixes.

Embodying Values

We can use technology to consciously and purposefully shape our ecotechnological world according to our wishes, if we realize that technology is complexly value laden and that we can embody our values in its creations. Unfortunately, conventional wisdom holds that technology is value free, so too many of us look elsewhere when seeking to find the source of undesirable artifacts and systems. Recently, however, social scientists have argued persuasively that technology, engineering, and science are value laden. They point out, for example, that military funding of both science and technology during the cold war decidedly influenced the problems that scientists and engineers chose to solve and the ideas and things that they created.

We readily acknowledge that architecture is value laden, and we should see that the artifacts and systems created by technology are as well. Monumental government buildings make us feel small and the government departments housed in them overwhelmingly powerful. The designs for a new Berlin prepared by Adolf Hitler's architect Albert Speer are a notorious example of an overwhelmingly domineering concept. The great French medieval cathedral at Chartres makes us seem insignificant in the presence of symbols expressing the power of the Almighty. Manhattan's skyline conveys the wealth of commerce. And the colleges at Cambridge University in England remind us of the great traditions of scholarship.

Architects have long argued that their buildings influence the behavior and health of people living and working in them. Between the two world wars, International Style architects in Germany designed housing developments with sunny kitchens, indoor toilets, sundecks, and gardens that workers from the inner cities had never known. Following in the steps of the great American landscape architect Frederick Law Olmsted, who designed Central Park in New York City, others designed green spaces in city centers reminding people that their environment need not express political power and commerce. Jane Jacobs, an urban reformer in 1961, asked planners to abandon their rational and sterile designs for redeveloping urban areas, so that people could mix and touch in long-standing, messy, vital city streets.

A closer look at engineering works and careful consideration of their impact upon us would remind us that technology is not value free. San Francisco's beautiful Golden Gate Bridge has become a city symbol that heightens civic pride. Parisians initially viewed the

Eiffel Tower with disdain because it expressed rational engineering values in a city that prided itself on its traditional culture. Now it stands as a towering symbol of French originality and imagination. Designers insist that customers buy automobiles to express the personal images that they wish to project. The redesigned small Volkswagen Beetle is a woman's car; the large Mercedes is a man's.

Not only architects and engineers embody values in their works, but users of technology do as well. Sometimes a device or system expresses the values of the users in unintended and unforeseen ways. An outstanding case in point is the heavy use of the Internet for e-mail correspondence, something that the designing engineers did not anticipate. Similarly, Alexander Graham Bell and other telephone pioneers expected the device to be utilized primarily for business transactions. The Wright brothers believed that the military would deploy the airplane simply for observation. And nineteenth-century developers of the internal combustion engine did not foresee its use in airplanes or automobiles; they expected it to displace small stationary steam engines.

Ecotechnological Systems

We should, therefore, be concerned about the values that we embody in our creations, especially our ecotechnological systems. Much of the planet consists of interacting natural and human-built systems, which together constitute ecotechnological systems. An architect taking into account natural forces, such as local climate, when designing a building is creating an ecotechnological system. Urban planners providing for the natural flow of rivers, streams, and prevailing winds are taking an ecological approach. Likewise, engineers and environmental scientists cooperating to restore ecological environments are often creating an ecotechnological system.

Shaped by both nature and technology, historic cities are ecotechnological. Settlers situated Philadelphia, Wilmington, Baltimore, Washington, Richmond, and other major eastern seaboard cities at the natural fall line of rivers, where the more easily eroded,

34. Nature stealthily shapes the human-built. Manhattan skyscrapers must cluster where the bedrock is substantial. Lower Manhattan, View Looking North (May 5, 1976), *by Thomas Airviews, box 28-32A, negative number 75010. Courtesy of the Collection of The New-York Historical Society.*

claylike soil of the tidewater region meets the more resistant, rock-strewn eastern border of the piedmont region. In an era of river navigation, cities at the fall line became places of trade, commerce, and portage where travelers and freight left boats for an overland passage around the falls. New York City starkly reveals natural circumstances, too. Builders located a cluster of skyscrapers in Midtown and another in the downtown financial district where bedrock, essential for foundations, runs near the surface. Sited to take advantage of natural features, great European historic cities are ecotechnological as well.

Today, however, cities often use technology to overwhelm nature rather than interact with it and adapt to it. Cities formerly developing in concert with nature are becoming simply human-built. Developers, for instance, fill in and divert streambeds, level hills, and fill in valleys in ways that do not allow water to flow naturally. Impervious paved surfaces on roads, parking lots, and roofs contribute to overly rapid and eroding storm-water runoff. Consequently, nature responds with impaired hydrological cycles that result in flooding. Improperly buried streams cause ground subsidence. High-rise buildings cut off the flow of pollution-clearing airflows and cause inversions that trap pollution from heavy traffic.

In *The Granite Garden* (1984), architect Anne Spirn contends that in the future cities should again respect natural features and provide an optimal compromise between nature and human purpose, thus creating a seamless ecotechnology. Fresh air, clean water, large and small urban parks, and the sustainable use of materials and energy should become commonplace.

Several cities in the United States and Europe have innovatively responded to the play of natural forces. Dayton, Ohio, found ways to counter the cold whipping winds of winter and the hot, humid air of summer that flowed into the city from the surrounding countryside. Using a six-foot-wide model of the city with buildings and streets to scale, graduate students in landscape design at Harvard University then studied the three-dimensional model in a wind tunnel. Drawing upon their studies, they proposed locating trees, new buildings, and streets to channel the flow of air from the surrounding countryside to remove polluted air, temper cold winds, and stimulate cooling breezes according to the season. Learning from this experiment, planners regulated new construction accordingly.

In Stuttgart, Germany, the city found that breezes cooling the city and carrying away polluted air came from surrounding hills, but that development there was diminishing the flow of air. Stuttgart prevented further development and located city parks to funnel air from the hills to cool and counter pollution. Other cities have designed parks with trees, grasses, shaded buildings, and water fountains, creating a summer oasis. In contrast, locating a park to receive the maximum sunshine can extend the pleasure of outdoors for several months in fall and early spring.

35. Architect Glenn Murcutt's designs take into account prevailing patterns of wind, rain, sun, and shade, as well as materials and client. Murcutt, Magney House, Bingie Bingie, South Coast NSW (1982–84). Photograph by Glenn Murcutt. Courtesy of G. Murcutt.

Water for drinking, plants, trees, cleaning, waste removal, and aesthetic enjoyment is an urban essential. Channeling water into a city has an ancient history, as Rome's eleven aqueducts testify. Urban rivers, streams, and storm water must be controlled in and around urban areas, however, to prevent flooding. In its management of water, Denver, Colorado, has shown the way for other cities. It transformed a rubble-strewn, flooding-prone downtown river into a "greenway" of parks, hiking areas, and bicycle paths. Inner-city plazas now help prevent flooding by storing storm water and then releasing it gradually.

Not only cities, but also houses can be ecotechnological. Glenn Murcutt, an Australian architect who won architecture's esteemed Pritzker Prize in 2002, designs place-related houses in a rational, minimalist manner. He defines a place in terms of water tables, rain and wind patterns, sun angles, soil types, and indigenous materials. Knowing the natural characteristics of a place through scientific

ecological analysis helps him to understand the possibilities of building in an appropriate manner, allowing natural features to shape
and interact with the human-built.

Murcutt's Magney House in Moruya, New South Wales, is representative of his designs. Extensive fenestration captures cool ocean
breezes in summer. In winter an insulated southern thermal glass
wall protects the house from cold winds while admitting the winter
sun. The earth on which the house rests warms it. Since there is no
groundwater at the site, the house collects rainwater in tanks. Outside walls are brick and metal clad, while glass opens the house to
the striking landscape. His architecture is called ecological func-
tionalism; it might be named ecotechnological.

Regions as well as buildings and cities can be ecotechnological.
In a book imaginatively titled *The Organic Machine* (1995), Richard
White, a Stanford University social and environmental historian,
writes of the long-standing interactive encounter between the forces
of nature and technology. He explores the creation of regional ecotechnological systems, or organic machines, that are seamless webs
of the natural (organic) and the human-built (machines). To develop
his theme, White focuses upon the Columbia River system in the
American Northwest, which has been transformed over time by numerous dams, channeling, and other human-built structures.

Native American fishermen, profit-driven developers, narrowly
focused dam builders, electric utility engineers, federal nuclear
bomb makers, and single-minded environmentalists have shaped
the Columbia River and it has shaped them. Because they have
transformed the river according to their various and contradictory
interests, the organic machine is inefficient. It is not a harmonious
and coherent marriage of the natural and human-built.

Similarly, environmental historian William Cronon convinc-
ingly describes in *Nature's Metropolis* (1991) how Chicago in the nineteenth century interacted with a hinterland extending from the Ap-
palachians to the Sierra Nevada in a mutually dependent way. In its
stockyards and mills, Chicago processed cattle and grains from the
farmlands and timber from virgin forests. It became the central

node in a commercial and railway system over which raw materials flowed into the metropolis and finished products out to a nation-wide market. The farmers and foresters depended upon Chicago for their income, as well as for manufactured products.

Kissimmee River Restoration Project

The federal government has financed ecotechnological projects in the past and present. Its use of technology to restore the vitality of a poverty-stricken Tennessee valley in the 1930s suggests the wide range of technology available to government in contrast to those at the disposal of private enterprise. The Tennessee Valley Author-ity (TVA), established by the administration of President Franklin Roosevelt, organized regional planners, idealistic engineers, and high-minded managers to preside over a multifaceted endeavor sys-tematically linking electric power generation, industrialization, soil revitalization, reforestation, navigation, and flood control.

David Lilienthal, a TVA commissioner, used a human-built world image when he observed that the country's progress no longer depended upon men armed with axes, rifles, and bowie knives, but upon teams of engineers, scientists, and administrators equipped with diesel engines, bulldozers, giant electric shovels, and modern organizational techniques. He stressed the moral responsibility of creators to respect the unity of nature, although environmental con-cerns, as defined today, played a minor role in the project. He also anticipated the future in calling for the participation of the people in technological decision making.

Today government is engaged in an ecotechnological project of comparable magnitude: the restoration of Florida's Kissimmee River system and the Everglades. From time immemorial, water flowed from a chain of lakes in the Kissimmee basin, north of Or-lando in central Florida, to Lake Kissimmee, south of Orlando. From Lake Kissimmee, the Kissimmee River, surrounded by a mile or two of floodplain, meandered southward for about a hundred

36. Kissimmee River in 1961 before channelization. Courtesy of the South Florida Water Management District.

miles to Lake Okeechobee. Thirty-five thousand acres of wetland covered the floodplain. From Lake Okeechobee, rainy-season flood-waters flowed in a great sheet, or "River of Grass," fifty miles wide and six inches deep to feed the extensive Everglades in southern-most Florida. Stretching almost two hundred miles, the Kissimmee-Everglades system supported an enormously diverse population of flora and fauna. Since 1962 this natural system has been violated for human purposes.

After a series of disastrous floods caused great damage to mush-rooming settlements in central and southern Florida, the U.S. Army Corps of Engineers between 1962 and 1971 constructed along the Kissimmee River system a massive network of canals, levees, dikes, and pumping stations to control floods and supply water to ex-panding population centers. Designed to carry off potential flood-waters in the most direct and quickest way, a straight canal nearly three hundred feet wide and fifty-six miles long replaced the for-

37. Kissimmee River after channelization. Courtesy of the South Florida Water Management District.

merly meandering Kissimmee River as the major drain to Lake Okeechobee.

Ecological consequences for the 3,000-square-mile Kissimmee River basin included the lowering of the water table, degradation of natural habitat, loss of wetlands, and transformation of former floodplains with their complex ecological systems into land for settlement and human cultivation, including cattle ranches. Because of the large amounts of phosphorous from animal wastes channeled directly into Lake Okeechobee, it became virtually a dead lake.

Stimulated by rising public concern for the environment and by

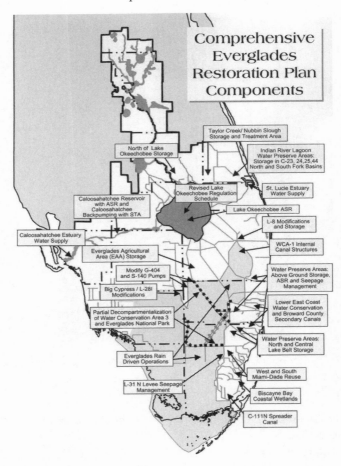

Comprehensive
Everglades
Restoration Plan
Components

Taylor Creek/ Nubbin Slough
Storage and Treatment Area

North of Lake
Okeechobee Storage

Indian River Lagoon
Water Preserve Areas:
Storage in C-23, 24,25,44
North and South Fork Basins

Revised Lake
Okeechobee Regulation
Schedule

St. Lucie Estuary
Water Supply

Caloosahatchee Reservoir
with ASR and
Caloosahatchee
Backpumping with STA

Lake Okeechobee ASR

L-8 Modifications
and Storage

Caloosahatchee Estuary
Water Supply

Everglades Agricultural
Area (EAA) Storage

WCA-1 Internal
Canal Structures

Modify G-404
and S-140 Pumps

Water Preserve Areas:
Above Ground Storage,
ASR and Seepage
Management

Big Cypress / L-28I
Modifications

Partial Decompartmentalization
of Water Conservation Area 3
and Everglades National Park

Lower East Coast
Water Conservation
and Broward County
Secondary Canals

Water Preserve Areas:
North and Central
Lake Belt Storage

Everglades Rain
Driven Operations

West and South
Miami-Dade Reuse

L-31 N Levee Seepage
Management

Biscayne Bay
Coastal Wetlands

C-111N Spreader
Canal

38. Components of the "Comprehensive Everglades Restoration Plan."

federal environmental legislation, in the 1970s federal and state agencies carried out feasibility studies and small-scale demonstration projects with the long-range goal of restoring the seasonal fluctuations of the Kissimmee River and lake as well as its floodplains and wetlands. In 1998 a ten-year, $600 million Army Corps of Engineers Kissimmee restoration project based upon these studies and demonstrations got under way to restore fifty-six miles of the upper reaches of the Kissimmee River and twenty-four thousand acres of its floodplain. Ecological scientists along with engineers have played a major role in the studies and in designing the restoration plan.

Everglades Restoration Plan

Besides channeling the Kissimmee, the U.S. Army Corps of Engineers had earlier also built levees, canals, and pump stations to move water around in the Everglades region to control flooding and supply water to where needed in the human-built world. Consequently, the Everglades suffered serious ecological damage as its water supply began to dry up. Taking this destruction into account as well as the need for flood control and water supply, the federal government and the state of Florida have agreed to fund the Everglades Restoration Plan, which will cost at least $7.8 billion.

Far more complex than the Kissimmee River project, the Everglades multipurpose restoration plan has been presented to the U.S. Congress and the public as essentially an ecological program designed to restore the Everglades by increasing the flow of water to it by removing engineering works and roads restricting the movement of the "River of Grass." The existing plan, which is subject to revision, calls for the Army Corps of Engineers in conjunction with its Florida state partner, the South Florida Water Management District, to begin in 2004 construction of 333 wells that will deep store water during flood times and pump it back where needed in dry periods. Eventually two giant limestone quarries will also store floodwaters that formally were channeled into the sea. The intent is to increase the flow of pollution-free water into the Everglades by one trillion gallons. The federal government and the state of Florida will share equally in the funding of the project.

Critics of the plan argue that the increased water supply will go mostly to agribusiness, industry, and the rapidly increasing south Florida population. Water for the Everglades will come later and in insufficient quantity to restore the Everglades. Critics also say that

Florida interests vested in development, commerce, industry, and agriculture will increasingly modify the ecological intentions of the original plan. In April 2003, for example, the sugarcane industry threatened to back away from its agreement to substantially reduce the phosphorus flowing from its fields into the Everglades by 2006, postponing the reduction by twenty years. Instead of a model project dedicated to ecological goals, it, according to critics, will end up a display of power by the politically privileged.

Yet, on the positive side, the planning has involved both engineers and environmental scientists. In the past, environmentalists and engineers often failed to cooperate in solving problems of the interacting natural and human-built worlds. Engineers dismissed environmentalists as tree huggers, and environmentalists labeled engineers as asphalt spreaders. Now, in the case of the Everglades, engineers and environmentalists have the opportunity to cooperate in the development of an ecotechnological project.

A Technological Fix?

Some environmentalists, however, feel that the Corps of Engineers, with its major technological plans for pumps and complex reservoirs, is still controlling nature instead of letting it find its own way. Others fear that the Corps of Engineers will preside over a "technological fix," as it has often done in the past and did earlier in the Kissimmee basin region. In simply replacing the meandering Kissimmee River with a straight canal and in constructing dikes and levees to control water flow, the corps ignored the complexity and variety of the environment-sustaining ecological system.

In a book titled *Seeing Like a State: How Certain Schemes to Improve the Human Condition Have Failed* (1998), James Scott argues that engineers, planners, and other professionals who are committed to applying science and mastering nature tend to reduce a complex, multivariable problem to an abstract, quantifiable simplification. In other words, they resort to technological fixes. They engage in reductionist practices because unwieldy reality then becomes legible, which allows for analysis and centralized management. Scott directs his attention primarily to schemes resulting in grossly flawed, highly planned modern cities, such as Brasília in Brazil, and to failed social engineering schemes of totalitarian revolutionaries, such as Stalinist collective agriculture in the 1930s.

Historical case studies make his point about reductionist approaches. He may be anticipating an Everglades project gone wrong with an opening chapter entitled "Nature and Space," which discusses scientific forestry in late-eighteenth-century Prussia and Saxony. Knowing that the regime looked narrowly at its forest through a fiscal lens focused upon revenue from timber, scientifically oriented state foresters launched a uniform plan of monoculture. They believed that the timber yield could be increased substantially by row planting Norwegian elms as a timber crop.

The reductionist focus upon timber harvest of coniferous trees ignored forests as sources of kindling, firewood, barks for medicine and tanning, leaves for fodder, and other things that peasants depended upon. The foresters, taking a rational, managerial approach, eliminated underbrush and cleared out flora and fauna of no interest to the state. After several decades, the initial harvest brought greater revenues than obtained from the former chaotic growth forest.

A great and unexpected loss, however, came during the second growth of the managed forest. Blight specific for the monoculture trees decimated the forest. Without the nutrient soil stemming from a symbiotic mix of underbrush, fallen branches and trees, insects, mammals, birds, and fungi, trees experienced stunted growth. The absence of diversity and complexity led to fragility and loss of resilience. In extreme cases, foresters encountered *Wildsterben* (forest death). Timber yields plummeted.

To drive home his point, Scott uses as an analogy a merchant who sends out a fleet of sailing ships of different designs and navigational equipment to avoid losing everything if disastrous storm conditions are encountered. Diversity saves the fleet. Highly planned, simplified policies lacking diversity and complexity face a fate not unlike the monoculture forest and the uniform fleet. Corps of Engineers take note!

Public Participation

Emphasizing diversity, the Kissimmee River restoration project offers a model for an ecotechnological approach with its alliance of engineers and environmental scientists who introduced complexity and diversity. The Kissimmee story also shows that in a democratic society, the political instruments are at hand enabling citizens, engineers, scientists, and ecologists to help shape their material world.

Since the federal environmental legislation enacted in 1969, government-funded public works require public participation in the design process. The system builders planning and designing the Boston Central Artery/ Tunnel (CA/ T) project, for example, took an enlightened, accepting approach to public participation and environmental regulations. In so doing, they moderated the resistance of the Boston activists who earlier blocked highway construction in downtown Boston.

Initiated in 1985 and scheduled for completion in 2004, the Central Artery/ Tunnel is the country's largest urban highway project and will cost more than $12 billion. CA/ T will replace a congested elevated highway that extends through the heart of inner-city Boston with an underground highway. It will also provide a new tunnel under Boston Harbor and a bridge across the Charles River to Cambridge. The project is designed to move traffic off the Boston streets and increase the flow of traffic from outside through Boston and to the airport.

Federal, state, and local governments, as well as numerous local interest groups, including those intent upon protecting ethnic neighborhood integrity, have had a voice in shaping the project. In accord with the law for federally funded highway projects, plans and designs have been made available in libraries and other public places. Public hearings have given organizations and individuals an

39. A handsome bridge design helped win over the Boston public to an enormously overbudget project. Bridge over the Charles River, courtesy of the Central Artery/Tunnel project.

opportunity to respond critically to the plans. For example, neighborhood activists have prevented the highway as well as access ramps and routes from physically dividing a neighborhood or overwhelming it with traffic. University professors and priests of the Catholic Church have played an influential role in stimulating and leading activists. And reporters on local newspapers, especially the *Boston Globe,* have energetically and effectively kept the public informed about project plans, especially those that threaten to have a negative impact upon the ecotechnological environment.

Citizens and organizations that treasure the Charles River flowing between Boston and Cambridge have successfully protected its banks from incursions and ensured that the new bridge across it will make a handsome statement about the region. Where CA/T has a negative impact upon the environment caused by the taking of parklands, for instance, it has mitigated the effects by creating compensating parklands. The project has funded various other mitigations that collectively have cost several billion dollars.

Project supporters predict that it will become a model for other cities throughout the world. Time will tell, however, if a systems approach in particular and engineering in general are moving away from a traditional centrally controlled, hierarchical style presided over by experts to a more complex, messy style where control is distributed, environmental concerns have a high priority, and the public is more involved. Those who oppose the new approach exemplified by CA/T call attention to the cost of public participation.

Technological Literacy

In order to participate effectively in project design specifically and technology policy generally, however, the public needs to learn about the engineering, architectural, and managerial processes used in creating and nurturing ecotechnological systems. The public needs to know who is making what decisions, so pressure can effectively be brought to bear to achieve its objectives. Environmentalists educate the public about endangered species, the decrease in biological diversity, and the loss of sustaining habitat. On the other hand, persons informing the public about technological change and the state of the ecotechnological environment, especially the human-built part of it, are few in number. To participate, the public needs to become technologically and aesthetically literate, as well as being morally concerned about the role of humans as creators.

For example, the public has experienced a steep learning curve and has participated effectively during the process leading up to the choice of a design concept for a building complex at the site of the destroyed Twin Towers in Lower Manhattan. Like CA/T, the design process has provided, so far, a model for public participation in the making of public spaces.

The process began not long after the Twin Towers fell to terrorists on September 11, 2001, when the Lower Manhattan Development Corporation oversaw an initial design competition on behalf of the Port Authority of New York and New Jersey, which owns the

site. The development corporation, aware of the symbolic significance of the designs, put them on display and invited a public reaction. The public, perhaps influenced by architectural critics commenting in the media, reacted so negatively that the development
corporation called for new conceptual designs from several of the
world's leading architectural firms. Six teams responded, four of
them collaborative proposals of individual firms. Newspapers, television, and websites carried illustrations and detailed accounts of
the designs during the succeeding months, and the development
corporation placed models on public display.

Public discussion ensued at a level that surprised architects and
critics. Herbert Muschamp, *New York Times* architectural critic,
commented that New Yorkers became involved in a major educa-
tional process about the program requirements of public spaces and
the relationship between social and artistic values. The public has
gained a new understanding of architecture and appreciates the
role that it can play in shaping it. Another commentator observed
that not since Gary Cooper played a heroic architect in the film
The Fountainhead had the public been so engrossed by architects and
architecture.

Realizing the role that public opinion would play, Daniel Libeskind, whose firm resides in Berlin, Germany, and the cooperative
team of Rafael Viñoly, Frederick Schwartz, Shigeru Ban, and Ken
Smith, the final two contestants in the competition, engaged in public relations ventures that brought them public appearances before
various audiences and on popular television programs. Such behavior was previously unknown in architectural circles, but public
participation was bringing major changes. Ultimately, public officials, the governor of New York State, and the mayor of New York
City were instrumental in choosing Libeskind.

As technology becomes increasing complex, however, effective
public participation in science and technology policy making may
become more difficult, even less likely. In "Why the Future Doesn't
Need Us" (2000), Bill Joy, a founder and chief scientist of Sun Mi-
crosystems, fears that complex, near-incomprehensible technology

endangers the human species. Allowing that genetic engineering, robotics, and nanotechnology may bring quality-of-life benefits, Joy, nevertheless, predicts that these technologies will contribute to a concentration of policy-making power in the hands of a very few knowledgeable scientists and engineers and the policy makers whom they advise. Awesomely complex technology may provide an opportunity for privileged elites to redesign not only the human-built world, but humans as well.

Not only will the public be ruled out of decision making by technological complexity, but the scientists and engineers who assume responsibility for the ecotechnological world have, Joy argues, arrogantly overestimated their ability to design the world well. He believes that the scientists who made atomic bombs to thwart National Socialist Germany did not anticipate Hiroshima, Nagasaki, and the cold war's horrendously dangerous arms race. Today, he reasons by analogy, engineers and scientists will not foresee where their creative activities, especially in genetic engineering, robotics, and nanotechnology, may lead in the civil and military realms. Yet others concerned about science and technology policy in a democracy believe that a much higher level of technological literacy might allow the public to effectively monitor and exercise control directly, or through their political representatives, over scientific and technological developments, even complex ones.

The successful lobbying and use of political power by organizations concerned about the natural environment offers hope that similar organizations will represent a technologically literate public concerned about the human-built and ecotechnological environments. Environmentalists and their organizations have campaigned for decades to educate the public through the media and books. Other organizations are beginning to promote technological literacy. The National Academy of Engineering, which speaks for leaders in the profession, recently sponsored a committee to study and issue a major report on technological literacy. Entitled *Technically Speaking: Why All Americans Need to Know More about Technology*, it calls for relevant education in schools, colleges, museums, the media, and elsewhere.

A technologically literate public might reject technological determinism and accept the current social science argument that technology is malleable and subject to social control. In recent years, historians and sociologists have provided many examples of socially constructed technology, the opposite of technological determinism. By socially constructed, they mean that the public, through organizations and as individuals, can make choices about the characteristics of the technology they use and the effects that it will have upon them. The Central Artery/Tunnel, for instance, is such a socially constructed project.

Today the endangered state of the natural environment, the deteriorating human-built world, and the threat of technology out of control reflect people's values and their resigning themselves to determinism. A change in values and an activist stance toward technological change will be an effective response to these pressing problems. Such a value change and activism will not come about, however, unless technology is better understood. This book is intended to provide such an understanding.

Bibliographic Essay

To understand technology in a historical perspective, I have drawn over the years upon the insights of historians, social scientists, scientists, engineers, philosophers, theologians, public intellectuals, and others. In the foregoing chapters, I have discussed the ideas in their books and articles that have made a lasting impression upon me. Assuming that my readers may wish to read at greater length in the sources that I have used, I am commenting in this brief bibliography upon the authors and their works, as well as publication details. I have organized the bibliography in categories reflecting the chapters of my book along with a section on defining technology.

- Defining Technology
- Second Creation
- Machine for Production
- Systems, Controls, and Information
- Technology and Culture
- Ecotechnological Environment

Defining Technology

Leo Marx, a historian at the Massachusetts Institute of Technology, and Eric Schatzberg, a historian at the University of Wisconsin, have helped me through dialogue and their publications to define technology. Troubled by the unreflective and casual ways in which historians have used the term, Marx has published several essays on the subject. Schatzberg has shared with me an essay, now in draft form, entitled "From Knowledge to Object: The Contested Meanings of Technology." Ed Constant, a historian of technology, has also engaged with me in an instructive dialogue about "technology."

Marx argues that "technology" is often used so broadly that it designates indiscriminately all the material resources that humans employ in making and doing. Consequentially, an author attempting to understand technology needs to define the context in which she or he is discussing technology. Technology takes meaning, for example, in the context of the historical period in which it is being considered. Marx has expressed his views on defining technology in several articles, including "Technology: The Emergence of a Hazardous Concept," *Social Research* 64 (Fall 1997): 965–88; "The Idea of 'Technology' and Postmodern Pessimism," in *Technology, Pessimism, and Postmodernism,* ed. Yaron Ezrahi, Everett Mendelsohn, and Howard P. Segal (Amherst: University of Massachusetts Press, 1995), pp. 11–25; and "American Literary Culture and the Fatalistic View of Technology," in *The Pilot and the Passenger: Essays on Literature, Technology, and Culture in the United States,* ed. Leo Marx (New York: Oxford University Press, 1988), pp. 127–38.

Schatzberg considers the varied definitions of technology over time. The two extremes are technology as knowledge of the practical arts and the practical arts themselves. He traces usage by Karl Marx, Thorstein Veblen, and Charles Beard, among others. Marx tended to define technology as principles governing production or as a field of knowledge. Veblen saw it as knowledge for productive purposes, especially of machine practice. And Beard associated it with relentless progress resulting from the mastery of the material environment by applying science.

Constant observes that technology is much more of an activity (or action) than artifacts. Economists like Joel Mokyr, who takes an evolutionary approach, view technology as a set of activities by which an organization accomplishes its economic tasks. The definition of technology in *International Encyclopedia of the Social Sciences*, ed. David L. Sills (New York: Macmillan, 1968), p. 557, suggests an activity as well: "technologies are bodies of skills, knowledge and procedure for making, using, and doing useful things." The English editors Charles Singer, E. J. Holmyard, A. R. Hall, and Trevor Williams of *A History of Technology*, 5 vols. (New York: Oxford University Press, 1954–58) define technology similarly as "how things are commonly done or made . . . [and] what things are done and made."

I have also drawn upon John Staudenmaier, editor of *Technology and Culture*, the journal of the Society for the History of Technology, who analyzes the use of "technology" in *Technology's Storytellers: Reweaving the Human Fabric* (Cambridge: MIT Press, 1985). Ruth Oldenziel in *Making Technology Masculine: Men, Women and Modern Machines in America, 1870–1945* (Amsterdam: Amsterdam University Press, 1999) explores the changing usage of "technology" and notes how today it implies a male professional activity. Ronald Kline has shown that this professional activity is associated with applied science in "Construing 'Technology' as 'Applied Science': Public Rhetoric of Scientists and Engineers in the United States, 1880–1945," *Isis* 86 (1995): 194–221.

My suggestion that Americans, in general, define technology too simply is supported by W. E. Dugger Jr. and L. C. Rose, "Itea/Gallup Poll Reveals What Americans Think about Technology" (Reston, Va.: International Technology Education Association, 2002).

Second Creation

Perry Miller's *The Life of the Mind in America: From the Revolution to the Civil War* (New York: Harcourt, Brace & World, 1965), a winner of the Pulitzer Prize, has served as a masterful example for me of the intellectual history of technology. Miller, a Harvard University professor of American literature, decidedly influenced the first generation of American studies scholars, including Leo Marx. The book has three major sections. The first is about evangelical Christianity and a great revival in the late eighteenth and early nineteenth centuries. The second deals with the rise of the legal profession in the United States during the first half of the nineteenth century from a disorganized, vaguely suspect group to a highly influential and respected profession with a systematic and logical approach to legal questions. In both, Miller writes like an angel, engaging his readers with wit, irony, and intellectual elegance.

Before his death in 1963 at the age of fifty-eight, Miller had not completed a third section, "Science—Theoretical and Applied." Elizabeth Miller, his wife, took pains to see that the completed first chapter for section three and outlined notes for the remainder reached publication along with the other two sections. The third section helped me to see the excitement that could be generated by a masterly approach to an intellectual history of American technology. Miller dismissed once and all for me the criticism that history of technology was nothing more than dull accounts of the progress of nuts and bolts. Miller created an identity for Americans as enthusiastic promoters of machines in technology's nation.

Elizabeth Miller graciously allowed me to see the notes that Perry Miller and his research assistants took when he was preparing to write book three. He imaginatively searched the middle ground, the conventional wisdom, looking for the mind of Americans. His sources included commencement day addresses, Fourth of July orations, and journal essays read by lawyers, amateur natural philosophers, literate farmers, and ambitious merchants. Miller had a genius for choosing the adjective that captured the mood of the writer or speaker. Sometime they were reprimanding; other times they were snarling, frequently platitudinous, and often self-congratulatory. Miller sampled like an opinion-poll taker today in order to define the American identity two centuries ago. My summary of his notes with his citations is with my papers, now housed at the archives of the Hagley Library in Wilmington, Delaware.

What I found in Miller that I have not encountered elsewhere is an understanding that Americans gave meaning in a religious con-

text to steamboats, locomotives, and factories—what we would la-
bel technological artifacts today. God, Americans believed, had instilled within them a divine spark that allowed them to complete his
creation. This is understandable among a people still viewing the
world through religious filters.

Yet the authors of Miller's sources often leave false impressions.
The settlers did not encounter a howling wilderness populated by a
few savages. They ventured into a landscape inhabited and partially
cultivated by over a million Native Americans. William Cronon, a
professor at the University of Wisconsin, in *Changes in the Land: Indians, Colonists, and the Ecology of New England* (New York: Hill and
Wang, 1983) has described the interactions of flora, fauna, Indians,
colonists, and natural forces in a complex and changing ecological
system within which social, political, and other events and trends
occurred. Carolyn Merchant in "Reinventing Eden: Western Culture as a Recovery Narrative," in *Uncommon Ground: Rethinking the
Human Place in Nature,* ed. William Cronon (New York: W. W. Norton, 1996), helped me understand the attitude of Native Americans
toward nature. Roderick Nash in his oft-reprinted *Wilderness and the
American Mind* (New Haven: Yale University Press, 1967) writes of
the changing concept of "wilderness" throughout history, and he
recognizes that an uncontrolled, threatening idea of wilderness has
been deeply rooted in the human mind since time immemorial.

I presented an early version of my thoughts about the second
creation in Thomas P. Hughes, "The Order of the Technological
World," in *History of Technology,* ed. A. Rupert Hall and Norman
Smith (London: Mansell, 1980), vol. 5, pp. 3–16, and in "The Second Creation of the World," in *Human Perspectives on Technology, Development, and Environment,* ed. Francis Sejersted and Ingunn Moser
(Oslo: Center for Technology and Culture, 1992), pp. 58–71.

MEPHISTOPHELEAN CREATIVITY

Like Miller's *Life of the Mind,* Johann Wolfgang von Goethe's *Faust,*
part II, act V, which he completed shortly before his death in 1832,
has shaped my concept of technology as a mode of creation. For a

reliable version of the poem, I have used Johann Wolfgang von Goethe, *Faust: Part Two*, translated from the German by Martin Greenberg (New Haven: Yale University Press, 1998). Loren Graham in *The Ghost of the Executed Engineer: Technology and the Fall of the Soviet Union* (Cambridge: Harvard University Press, 1993) describes recent project builders whose behavior is not unlike the ruthlessness of Faust the creator. Ronald Gray's "Goethe's *Faust*, Part II," *Cambridge Quarterly* I (Winter 1965–66): 355–79, influenced my interpretation of the myth of creation, as has Jaroslav Pelikan's *Faust the Theologian* (New Haven: Yale University Press, 1995).

MECHANICAL ARTS

Although my focus is upon the attitudes of Americans creating the human-built world during the early nineteenth century, I also consider their European ideological heritage pertaining to technology and creation that gave meaning to their venture. Among helpful sources are Friedrich Klemm, *A History of Western Technology* (New York: Charles Scribner, 1959); Martha Teach Gnudi and Eugene Ferguson, who translated and wrote an introduction to one of the most handsome of Renaissance machine books: Agostino Ramelli's *The Various and Ingenious Machines of Captain Agostino Ramelli*, first published in 1588 (Baltimore: Johns Hopkins University Press, 1976); and Clarence J. Glacken, *Traces on the Rhodian Shore: Nature and Culture in Western Thought from Ancient Times to the End of the Eighteenth Century* (Berkeley: University of California Press, 1967). I am indebted to professor Erik Swyngedouw of Oxford University for bringing Glacken's work to my attention.

TECHNOLOGICAL MILLENNIUM

In placing the settlers' heritage in a religious, especially Puritan, context, I have drawn upon Charles Webster's *The Great Instauration: Science, Medicine and Reform, 1626–1660* (London: Gerald Duckworth, 1975). David Noble has also interestingly explored the millenarian theme pertaining to the Edenic recovery in *The Religion of Technology: The Divinity of Man and the Spirit of Invention* (New York: Alfred A. Knopf, 1997). Robert Merton in his 1938 dissertation, later published as *Science, Technology & Society in Seventeenth-Century England* (New York: Howard Fertig, 1993), documents thoughtfully the ways in which the Puritan ethic cultivated the study of science and its application, an approach carried over by English settlers to America.

PARADISE REGAINED

A vision of an American golden age that could be ushered in by the mechanical arts is represented by J. A. (John Adolphus) Etzler, *The Paradise within the Reach of All Men: Without Labour, by Powers of Nature and Machinery: An Address to All Intelligent Men* (London: John Brooks, 1836). It caught the attention of Henry David Thoreau, who responded to Etzler in an essay entitled "Paradise (to Be) Regained," which appeared in the *United States Magazine and Democratic Review* 13 (November 1843). Thoreau, like many who decry consumerism today, rejoined that only matters of the spirit truly fulfilled human longing. Historian Brooke Hindle in *Technology in Early America* (Chapel Hill: University of North Carolina Press, 1966) further explores Americans' thrill in the technological transformation.

MACHINE IN THE GARDEN AND TECHNOLOGICAL SUBLIME

Leo Marx's *The Machine in the Garden: Technology and the Pastoral Ideal in America* (New York: Oxford University Press, 1964) offers an imaginative and engagingly written account of American attitudes toward technology and locates these centrally in the nation's history. While relying primarily upon major literary sources, such as the writings of Nathaniel Hawthorne, Herman Melville, and Ralph Waldo Emerson, Marx suggests that this literature reflected the opinions of society at large. I, like many other historians of technology, eager to move our subject matter from the periphery toward the center of the stage of history, happily placed *The Machine in the*

Garden in the hands of my students. I found Marx's presentation of early-nineteenth-century lawyer Timothy Walker's argument that machinery would bring intellectual and social progress, as well as economic, more central to my teaching than Alexander Hamilton's 1791 *Report on the Subject of Manufactures,* which is grist for the mill of economic historians.

Thomas Carlyle's essay "Signs of the Times," *Edinburgh Review* 98 (1829), which was highly critical of the mechanization of the world, stimulated Walker's response. Marx's views on technology and nature can also be found in Marx, "Does Pastoralism Have a Future," in *The Pastoral Landscape: Studies in the History of Art* (Washington, D.C.: Center for the Advanced Study in the Visual Arts, National Gallery of Art, 1992), pp. 209–23.

As Jeffrey L. Meikle has shown in "Classics Revisited: Leo Marx's *Machine in the Garden,*" *Technology and Culture* 44 (January 2003): 147–59, other scholars have followed the ramifications of Marx's thesis. John F. Kasson in *Civilizing the Machine: Technology and Republican Values in America, 1776–1900* (New York: Penguin, 1977) has found in popular writing American concern for the technological transformation. David Nye has engagingly explored Americans' awestruck reactions to the power of technology in *American Technological Sublime* (Cambridge: MIT Press, 1994).

JEFFERSON AND EMERSON

In seeking meanings attached to the creation of a human-built world by a number of prominent, especially literary, figures in nineteenth-century America, I have focused upon Thomas Jefferson and Ralph Waldo Emerson. For Jefferson, I have relied upon Dumas Malone, *Jefferson, the Virginian* (Boston: Little, Brown, 1948); *Jefferson Himself: The Personal Narrative of a Many-Sided American,* ed. Bernard Mayo (Charlottesville: University Press of Virginia, 1942); Edwin Martin, *Thomas Jefferson: Scientist* (New York: Collier, 1961); and Joseph Ellis, *American Sphinx: The Character of Thomas Jefferson* (New York: Vintage, 1998). For Emerson's views, I have used several of his essays including "Nature" and "Wealth" in *The Selected Writ-*

ings of Ralph Waldo Emerson, ed. Brooks Atkinson (New York: Modern Library, 1992).

Instead of a gardenlike Edenic recovery, Americans fostered a nature-exploiting machine for production. Lewis Mumford in *The City in History: Its Origins, Its Transformations, and Its Prospects* (New York: Harcourt, Brace & World, 1961) describes the dank, dark late-nineteenth-century American industrial cities, the reality of which dashed hopes of those who earlier envisaged a pastoral America. The egregious case was Pittsburgh as described in *City at the Point: Essays on the Social History of Pittsburgh,* ed. Samuel Hays (Pittsburgh: University of Pittsburgh Press, 1989), and in *Pittsburgh Surveyed: Social Science and Social Reform in the Early Twentieth Century,* ed. Maurine W. Greenwald and Margo Anderson (Pittsburgh: University of Pittsburgh Press, 1996). I am also indebted to David McCullough, who allowed me to see his unpublished essay about Pittsburgh, "Hell with the Lid Off."

The horrendous wasting of a dreamed-of Edenic landscape is described in the dissertations of Fredric Lincoln Quivik, "Smoke and Tailings: An Environmental History of Copper Smelting Technologies in Montana, 1880–1930" (University of Pennsylvania, 1998), and Timothy J. Lecain, "Moving Mountains: Technology and the Environment in Western Copper Mining" (University of Delaware, 1998). On Lynn White's view of our ecological crises, see his "The Historic Roots of Our Ecological Crisis," *Science* (March 10, 1967).

Machine for Production

THE SECOND INDUSTRIAL REVOLUTION

A second industrial revolution stimulated Americans' and Germans' fascination with machines. In *American Genesis: A Century of In-*

vention and Technological Enthusiasm, 1870–1970 (New York: Viking, 1989), I have described the technological, organizational, and cultural changes defining this revolution. Among its hallmarks was the Ford system of mass production, which has been well explained in David A. Hounshell, *From the American System to Mass Production, 1800–1932: The Development of Manufacturing Technology in the United States* (Baltimore: Johns Hopkins University Press, 1984). The electrification of Germany and the United States also characterized the revolution, and I have discussed this in *Networks of Power: Electrification in Western Society, 1880–1930* (Baltimore: Johns Hopkins University Press, 1983). In his autobiography *The Education of Henry Adams* (1918), Adams captures the remarkable way in which electricity charged the public imagination in a chapter memorably titled "Dynamo and the Virgin." While I have only mentioned in passing the Soviet Union's efforts to adopt and adapt the technology of the second industrial revolution, more about this can be found in Paul R. Josephson, *Totalitarian Science and Technology, Control of Nature* (Atlantic Highlands, N.J.: Humanities Books, 1996), and in my *American Genesis.*

MODERN METROPOLIS

Revolutionary technological, social, organizational, and cultural changes were obvious in industrial metropolises. Still a seminal source for the subject is Lewis Mumford, *The City in History: Its Origins, Its Transformations, and Its Prospects* (New York: Harcourt, Brace & World, 1961). The transformation of New York City during the late nineteenth and early twentieth centuries is described in Carl Condit, *The Port of New York: A History of the Rail and Terminal System from the Beginnings to Pennsylvania Station* (Chicago: University of Chicago Press, 1980). I have explored ways in which the second industrial revolution took shape in, and shaped, Berlin and New York City in "The City as Creator and Creation," in *Berlin / New York: Like and Unlike: Essays on Architecture and Art from 1870 to the Present*, ed. Josef Paul Kleihues and Christina Rathgeber (New York: Rizzoli, 1993), pp. 13–32.

CONTROL

Americans' and Germans' enthusiasm for machines can be explained partially by their confidence that they could control them. This confidence stemmed from the appreciable success that engineers were having in inventing and developing feedback controls for mechanical and electrical systems. A modern history of controls can be found in my biography of *Elmer Sperry: Inventor and Engineer* (1971; reprint, Baltimore: Johns Hopkins Press, 1993). David Mindell has written a detailed history of twentieth-century controls in *Between Human and Machine: Feedback, Control, and Computing Before Cybernetics* (Baltimore: Johns Hopkins University Press, 2002). The coupling of technology with control is discussed by several authors in *Cultures of Control*, ed. Miriam Levin (Amsterdam: Harwood Academic Publishers, 2000).

SPENGLER

I find that the reactions of German historian Oswald Spengler to modern machine technology contrast dramatically with the attitudes of many of his contemporaries, as well as ours today. He considered the centrality of an ideology of engineering in the modern West a sign that it was losing the values associated with art, philosophy, and high culture. As a result, the West was becoming a prideful civilization rather than a culture. His views can be found in his widely read *The Decline of the West*, trans. Charles Francis Atkinson (New York: Alfred A. Knopf, 1939). See especially chapter 14, entitled "The Machine."

In 1931 Spengler published a small book in which he particularized his views on technology in history. It has been translated as *Man and Technics: A Contribution to the Philosophy of Life*, trans. Charles Francis Atkinson (New York: Alfred Knopf, 1963). H. Stuart Hughes, a cultural historian, dismisses the book as superficial in *Oswald Spengler: A Critical Estimate* (New York: Scribner, 1962). On the other hand, John Farrenkopf in *Prophet of Decline: Spengler on World History and Politics* (Baton Rouge: Louisiana State University Press, 2001) finds *Man and Technics* to be a gem of a book that reveals Spengler's insightful and prescient views on technology, especially its relationship with nature. Farrenkopf has had much greater access to Spengler papers than Hughes.

I have written about the ideology of Nazi engineers in a way suggestive of Spengler's views in "Technology," in *The Holocaust,* ed. Henry Friedlander and Sybil Milton (Millwood, N.Y.: Kraus, 1980), pp. 165–81.

MUMFORD

Lewis Mumford, an articulate and insightful critic of technology's impact upon society, published his broad-ranging history of technology in *Technics and Civilization* (New York: Harcourt, Brace and Co., 1934). Donald Miller's biography of Mumford, *Lewis Mumford: A Life* (New York: Weidenfeld & Nicolson, 1989), deals not only with the man, but his numerous books and articles. Mumford's concern that the spirit of mechanization would overwhelm the organic culture that he valued is discussed by Leo Marx in "Lewis Mumford: Prophet of Organicism," in *Lewis Mumford: Public Intellectual,* ed. Thomas P. Hughes and Agatha C. Hughes (New York: Oxford University Press, 1990), pp. 164–80. The Hughes and Hughes volume has essays by informed scholars on many facets of Mumford's philosophy of technology, including Rosalind William's "Lewis Mumford as a Historian of Technology in *Technics and Civilization,*" pp. 43–65, and an introduction by the Hugheses, pp. 3–19. Williams has written "Classics Revisited: Lewis Mumford's *Technics and Civilization,*" *Technology and Culture* 43, no. 1 (2002): 139–49.

SOMBART

Werner Sombart, a major German economic historian of the early twentieth century, believed technology to be a major shaper of history. He did not share the deep pessimism of Spengler about technology and culture, but took an ambivalent stance. He rejected

Karl Marx's technological determinism, seeing, instead, technology as one of many causal factors shaping the evolution of society. For an analysis of Marx's attitude toward technology, see Donald MacKenzie, "Marx and the Machine," *Technology and Culture* 25 (1984): 473–502.

Sombart discusses technology's interaction with other forces shaping history in an essay entitled "The Influence of Technical Inventions," which is reprinted in Werner Sombart, *Economic Life in the Modern Age*, ed. Nico Stehr and Reiner Grundman (New Brunswick: Transaction Publishers, 2001). In the introduction, the editors provide autobiographical material on Sombart and a survey of his major themes and methodology. Sombart also expresses his views on technology in "Technik und Kultur," *Archiv für Socialwissenschaft und Socialpolitik* 33 (1911): 305–47. Sombart's major work is *Der moderne Kapitalismus historisch-systematische Darstellung des gesamteuropaischen Wirtschaftslebens von seinen Anfangen bis zur Gegenwart* (München: Duncker & Humblot, 1919), 3 volumes.

RATHENAU

Walther Rathenau deeply felt and persuasively expressed the complexity and contradictions of the modern Western world as it took shape early in the twentieth century. Mumford probably took some of his views about the tension between the mechanical and the organic from Rathenau. Rathenau reached a literate German public with his books about modern culture. Harry Graf Kessler provides access to his views in *Walther Rathenau: His Life and Work* (New York: Howard Fertig, 1969). Several of Rathenau's books have been translated into English: *In Days to Come*, trans. Eden and Cedar Paul (London: G. Allen & Unwin, 1921), and *The New Society*, trans. Arthur Windham (New York: Harcourt, Brace & Howe, 1921). His notes and diaries, 1907–1922, have been published as *Walther Rathenau: Industrialist, Banker, Intellectual, and Politician*, ed. Hartmut Pogge von Strandmann, trans. Caroline Pinder-Cracraft in conjunction with Hilary and Hartmut Pogge von Strandmann (Oxford: Clarendon Press, 1985).

I have written about Rathenau's ambiguous attitudes toward the rational and the organic in "Walther Rathenau: System Builder," in *Ein Mann vieler Eigenschaften: Walter Rathenau und die Kultur der Moderne*, ed. T. Hughes, T. Buddensieg, and J. Kocka (Berlin: Wagenbach, 1990), pp. 9–31. A negative analysis of Rathenau is found in David Felix, "Walther Rathenau: The Bad Thinker and His Uses," *European Studies Review* 5 (1975): 69–79. See also Claudia Ann Kounz, "Walther Rathenau's Vision of the Future: The Etiology of an Ideal" (Ph.D. diss., Rutgers University, 1969), and Peter Jacob Loewenberg, "Walther Rathenau and German Society," (Ph.D. diss., University of California, Berkeley, 1966). A fascinating interpretation of Rathenau can be found in the fictional person of Paul Arnheim portrayed in Robert Musil's great novel *The Man without Qualities*, trans. Sophie Wilkins (New York: Alfred Knopf, 1995).

BEARD

The contrast between the ambiguous and subtlety nuanced attitude of Rathenau and the uncritical enthusiasm of Charles Beard for technology is dramatic. While Rathenau expressed the doubts of a European intellectual about the impact of machine technology on high culture, Beard celebrated Americans' conviction that their inventors and engineers were leading them into a bright future filled with consumer goods. This point of view can be found in Beard and his wife's coauthored, enormously successful text *The Rise of American Civilization: The Industrial Era* (New York: Macmillan, 1927). I have drawn upon the chapter entitled "The Machine." It is an impressive cultural and social survey of America during the machine age, which extended from about 1880 to 1930.

I have also relied on Beard's introduction to a book he edited entitled *Toward Civilization* (New York: Longmans, Green, 1930). In it, he praises American engineers and also castigates European aesthetes who do not appreciate the American achievement. Among those criticized is Richard Müller-Freienfels, who wrote *Mysteries of the Soul* (London: G. Allen & Unwin, 1929). Beard's general views on

history can be found in Richard Hofstadter, *The Progressive Historians: Turner, Beard, Parrington* (Chicago: University of Chicago Press,
1979). Beard contemporaries who shared his appreciation of the
American achievement include Arthur H. Compton, whose views
are expressed in "Oxford and Chicago: A Contrast," *Scribner's Magazine* 99 (1936): 355ff., and Matthew Josephson in "Made in America," *Broom* 2 (June 1922): 269–70.

Systems, Controls, and Information

SYSTEMS ERA

Further discussion of the systems era can be found in chapter 4
of my *Rescuing Prometheus* (New York: Pantheon Books, 1998). For
Russell Ackoff's criticism of the systems approach, see *Rescuing
Prometheus*. Gene Rochlin, Todd La Porte, and Karlene H. Roberts
explore the modes of managing super-complex systems in "The
Self-Designing High-Reliability Organization: Aircraft Carrier
Flight Operations at Sea," *Naval War College Review* (Autumn 1987):
76–90. More on the Taylorist approach to management can be
found in Frederick W. Taylor, *The Principles of Scientific Management*
(New York: Harper & Brothers, 1911).

WEAPONS SYSTEMS

Merritt Roe Smith and others write about ways in which the military has shaped civil management in the United States throughout
its history in *Military Enterprise and Technological Change: Perspectives on
the American Experience,* ed. Merritt Roe Smith (Cambridge: MIT
Press, 1985).

For the use of the systems approach during World War II, see
David A. Mindell, "Automation's Finest Hour: Radar and System
Integration in World War II," pp. 27–56, and Erik P. Rau, "The
Adoption of Operations Research in the United States during

World War II," pp. 57–92, both in *Systems, Experts and Computers,* ed. Agatha C. Hughes and Thomas P. Hughes (Cambridge: MIT Press, 2000).

For more on the SAGE aircraft defense system, see Hughes, *Rescuing Prometheus.* The spread of the SAGE management approach throughout the U.S. Air Force is discussed in John F. Jacobs, *Air Force Command and Control System Development* (Bedford, Mass.: MITRE Corporation, 1961). In writing the summary of the Polaris weapons project, I have drawn upon Erik Rau, "Polaris," an unpublished essay, University of Pennsylvania, 1993. Glenn Bugos explains the nature of the systems approach to weapons development in *Engineering the F-4 Phantom II: Parts into Systems* (Annapolis: Naval Institute Press, 1996). On the management of weapons systems, see also Stephen B. Johnson, "From Concurrency to Phased Planning: An Episode in the History of Systems Management," in *Systems, Experts, and Computers,* ed. Hughes and Hughes, pp. 93–112.

THE MILITARY-INDUSTRIAL-UNIVERSITY COMPLEX

Seymour Melman describes the rise of the military-industrial-university complex in *The Permanent War Economy: American Capitalism in Decline* (New York: Simon and Schuster, 1985) and *Pentagon Capitalism: The Political Economy of War* (San Francisco: McGraw-Hill, 1970). Paul Forman shows how the complex influenced the course of science development in "Behind Quantum Electronics: National Security as Basis for Physical Research in the United States, 1940–1960," *Historical Studies in the Physical and Biological Sciences* 18, no. 1 (1987): 149–229.

SPREAD OF THE SYSTEMS APPROACH

Atsushi Akera, Fred Quivik, and Erik Rau of the University of Pennsylvania provided research assistance in the preparation of this section on the spread of a systems approach. I am also indebted to Elliot Fishman for research. Other essays on the spread of the systems approach from the military to the civilian section include the following essays in *Systems, Experts, and Computers*, ed. Hughes and Hughes: Glenn Bugos, "System Reshapes the Corporation: Joint Ventures in the Bay Area Rapid Transit System, 1962–1972," pp. 113–32; David R. Jardini, "Out of the Blue Yonder: The Transfer of Systems Thinking from the Pentagon to the Great Society, 1961–1965," pp. 311–58; Davis Dyer, "The Limits of Technology Transfer: Civil Systems at TRW, 1965–1975," pp. 359–84; Arne Kaijser and Joar Tiberg, "From Operations Research to Futures Studies: The Establishment, Diffusion, and Transformation of the Systems Approach in Sweden, 1945–1980," pp. 385–412; Harvey Brooks and Alan McDonald, "The International Institute for Applied Systems Analysis," pp. 413–32; Roger Levien, "RAND, IIASA, and the Conduct of Systems Analysis," pp. 433–62; Gabrielle Hecht, "Planning a Technological Nation: Systems Thinking and Politics of National Identity in Postwar France," pp. 133–60; and Paul Edwards, "The World in a Machine: Origins and Impacts of Early Computerized Global System Models," pp. 221–54.

URBAN SYSTEMS: PROMISE AND FAILURE

Robert A. Caro in *The Power Broker: Robert Moses and the Fall of New York* (New York: Vintage, 1975) finds the older Moses autocratic and destructive of the fabric of urban life. Moses refutes his critics, especially Caro, in an unpublished essay "Document on a *New Yorker* Profile and Biography," August 26, 1974. I am indebted to Prof. Dr. Bernward Joerges of the *Wissenschaftszentrum* Berlin for a copy of this essay. For more on Moses, see *Robert Moses: Single-Minded Genius*, ed. Joann P. Krieg (Interlaken, N.Y.: Heart of the Lakes Publishing, 1989).

On the urban infrastructure, see *Technology and the Rise of the Networked City in Europe and North America*, ed. Joel Tarr and G. Dupuy (Philadelphia: Temple University Press, 1998); Stephen Graham and Simon Marvin, *Splintering Urbanism: Networked Infrastructures, Technological Mobilities, and the Urban Condition* (London: Routledge,

2001); and Rai Zimmerman, "Social Implications of Infrastructure Network Interactions," paper presented at the Social Sustainability of Technological Networks conference, New York, N.Y., April 17–21, 2001.

SYSTEMS DISCREDITED: THE VIETNAM WAR AND THE COUNTERCULTURE

By the time Lewis Mumford wrote *The Pentagon of Power: The Myth of the Machine* (New York: Harcourt Brace Jovanovich, 1970), his anxiety about the spread of a megamachine became for him a near obsession. He saw the megamachine, or a large sociotechnological weapons system, as shaping the course of modern history and likely to bring civilization to a violent end in a nuclear holocaust. *Lewis Mumford: Public Intellectual,* ed. Hughes and Hughes, has essays about Mumford's megasystem anxiety by Donald I. Miller, "The Myth of the Machine: Technics and Human Development," pp. 152–64; and Everett Mendelsohn, "Prophet of Our Discontent: Lewis Mumford Confronts the Bomb," pp. 343–60.

Elting Morison, a historian at the Massachusetts Institute of Technology, and Perry Miller, whom we encountered above, both feared that technological systems running out of control were threatening to determine the course of human history. Their views can be found in Miller, "The Responsibility of Mind in a Civilization of Machines," in *The Responsibility of Mind in a Civilization of Machines* (Amherst: University of Massachusetts Press, 1979), pp. 65–83; and in Morison, *From Know-How to Nowhere* (New York: Basic Books, 1974), and *Men, Machines, and Modern Times* (Cambridge: MIT Press, 1984).

Among negative reactions to the use of weapons systems during the Vietnam War, John McDermott's "Technology: The Opiate of the Intellectuals," *New York Review of Books* 13 (July 31, 1969): 25–35, is especially memorable. Counterculture critics also lamented the impact of large technological systems upon the environment. Their works include Rachel Carson, *Silent Spring* (Boston: Houghton

Mifflin, 1962); Barry Commoner, *The Closing Circle; Nature, Man, and Technology* (New York: Knopf, 1971); Amory B. Lovins, *Soft Energy Paths: Toward a Durable Peace* (New York: Harper & Row, 1979); and *The Whole Earth Catalog,* ed. Stewart Brand et al. (Menlo Park, Calif.: Nowels, 1968).

CONTROLS AND INFORMATION

Machines seemed subject to control, but systems were another matter. The systems era stimulated the publication of academic and popular books by scientists, engineers, and social scientists about controls. Influential among them was Norbert Wiener's *Cybernetics; or, Control and Communication in the Animal and the Machine* (Cambridge: MIT Press, 1948). Wiener took "cybernetics" from the Greek *xebernetis,* meaning helmsman, which suggests the inclusive character of his concepts. Coupled with control is the subject of communication, a theory about which is presented and explained by Claude E. Shannon and Warren Weaver in *The Mathematical Theory of Communication* (Urbana: University of Illinois Press, 1949).

Feedback controls have a long history; see, for example, Otto Mayr, *Liberty, Authority, and Automatic Machinery in Early Modern Europe* (Baltimore: Johns Hopkins University Press, 1993); and Thomas Parke Hughes, *Elmer Sperry: Inventor and Engineer.* Recently David Mindell has written a detailed history of controls in *Between Human and Machine: Feedback, Control, and Computing Before Cybernetics.*

Concepts associated with cybernetics spread into other disciplines, as indicated by Arturo Rosenblueth, Norbert Wiener, and Julian Bigelow, "Behavior, Purpose, and Teleology," *Philosophy of Science* 10, no. 1 (1943): 18–24. An account of the activities and publications of the enthusiastic followers of Wiener is engagingly given in Steve Heims, *The Cybernetics Group* (Cambridge: MIT Press, 1991). Heims has also written about Wiener and another expert on controls and feedback in *John Von Neumann and Norbert Wiener: From Mathematics to the Technologies of Life and Death* (Cambridge: MIT Press, 1980).

FLOOD OF INFORMATION METAPHORS:
MOLECULAR AND DEVELOPMENTAL BIOLOGY

Wiener and other cyberneticians influenced developmental biologists, whose theory of development was heavily laced with information and control metaphors. Their acceptance of a cybernetics approach signaled a growing interest among scientists and social scientists in information theory, an interest that was a harbinger of the nascent information revolution.

Evelyn Fox Keller explained to me the different approaches of developmental and molecular biologists. Her *Refiguring Life: Metaphors of Twentieth-Century Biology* (New York: Columbia University Press, 1995) is a superb introduction to recent biology. Lily E. Kay's "How a Genetic Code Became an Information System," in *Systems, Experts, and Computers*, ed. Hughes and Hughes, pp. 463–92, further illuminates the relationship between new approaches in biology and information. On James Watson, see Victor K. McElheny, *Watson and DNA: Making a Scientific Revolution* (Cambridge, Mass.: Perseus Books, 2003).

THE INFORMATION REVOLUTION

Other indications of a forthcoming information revolution grew after World War II. Derek J. de Solla Price in *Little Science, Big Science* (New York: Columbia University Press, 1963) observes that the number of scientists and science citations was increasing exponentially. G. Pascal Zachary in *Endless Frontier: Vannevar Bush, Engineer of the American Century* (New York: Free Press, 1997) recalls that during World War II Bush found himself overwhelmed by information and craving machines to organize and sort it. James R. Beniger alerted the scholarly world to the arrival of an information society in his prescient *The Control Revolution: Technological and Economic Origins of the Information Society* (Cambridge: Harvard University Press, 1986). Recently Joel Mokyr has traced the origins of a knowledge economy back to earlier roots in *The Gifts of Athena: Historical Origins of the*

Knowledge Economy (Princeton: Princeton University Press, 2003), as have Alfred D. Chandler and James W. Cortada, eds., in *A Nation Transformed by Information: How Information Has Shaped the United States from Colonial Times to the Present* (New York: Oxford University Press, 2000).

For information about the technical core of the information revolution, I am indebted to Paul Ceruzzi, curator at the Smithsonian National Air and Space Museum, and Paul Edwards, professor at the University of Michigan. Informative books on the subject have been written by Michael Riordan and Lillian Hoddeson, authors of *Crystal Fire: The Birth of the Information Age* (New York: W. W. Norton, 1997); AnnaLee Saxenian, *Regional Advantage: Culture and Competition in Silicon Valley and Route 128* (Cambridge: Harvard University Press, 1996); Steven Levy, *Hackers: Heroes of the Computer Revolution* (Garden City, N.Y.: Anchor Press/Doubleday, 1984); and Manuel Castells, *The Rise of the Network Society* (Oxford, Eng.: Blackwell, 1996). On John Bardeen, see Lillian Hoddeson and Vicki Daitch, *True Genius: The Life and Science of John Bardeen* (Washington, D.C.: Joseph Henry Press, 2002).

On the computer, see Paul Ceruzzi, *A History of Modern Computing* (Cambridge: MIT Press, 1998). For the Internet, see Janet Abbate, *Inventing the Internet* (Cambridge: MIT Press, 1999). For an overview of books and articles on the information age, see Richard R. John, "Rendezvous with Information? Computers and Communications Networks in the United States," *Business History Review* 75 (2001): 1–13.

REACTIONS TO THE INFORMATION REVOLUTION

In surveying reactions to the information revolution, I have drawn upon Chris Abel, *Architecture and Identity: Towards a Global Eco-Culture* (Oxford, Eng.: Architectural Press, 1997); Castells, *The Rise of the Network Society;* Paul Edwards, *The Closed World: Computers and the Politics of Discourse in Cold War America* (Cambridge: MIT Press, 1996); Tom Forester, *High-Tech Society: The Story of the Information Technology Revolution* (Cambridge: MIT Press, 1987); Bill Gates with

Nathan Myhrvold and Peter Rinearson, *The Road Ahead* (New York: Viking, 1995); and Nicholas Negroponte, *Being Digital* (New York: Alfred A. Knopf, 1995). I have also used my own "Ort, Raum, Technik und Architektur," in *Die Architektur, die Tradition und der Ort*, ed. Vittorio Magnago Lampugnani (München: Anstalt, 2000), as well as Torsten Hagerstrand, "Aspects of the Spatial Structure of Social Communication and the Diffusion of Information," in *Man, Space and Environment* (New York: Oxford University Press, 1972), pp. 328–40.

Technology and Culture

Robert Hughes provides a provocative and insightful introduction to modern art and architecture in *The Shock of the New* (New York: Alfred Knopf, 1981). He surveys American art in *American Visions: The Epic History of Art in America* (New York: Alfred Knopf, 1997). Also see Richard Guy Wilson, Dianne H. Pilgrim, and Dickran Tashjian, *The Machine Age in America, 1918–1941* (New York: Harry N. Abrams, 1986). I have also discussed the ways in which American technology, notably Fordism and Taylorism, influenced modern European art and architecture in *American Genesis*. The pre–World War I alliance between German industry, the arts, and the crafts is detailed in Joan Campbell, *The German Werkbund* (Princeton: Princeton University Press, 1978). The German architect Hermann Muthesius encouraged German industry to design products of beauty, but especially suited for mass production, in "Handarbeit und Massenerzeugnis," *Technischer Abend im Zentralinstitut für Erziehung und Unterricht* 4 (1917).

BEHRENS

Peter Behrens expressed in design and architecture German enthusiasm for machine symbols and machine production. His work has been admirably analyzed and set in historical perspective in the introduction by Tilmann Buddensieg to *Peter Behrens and the AEG, 1907–1914,* ed. Tilmann Buddensieg (Cambridge: MIT Press, 1984). Among the informative articles in the Buddensieg volume is Karen Wilhelm, "Fabrikenkunst: The Turbine Hall and What Came of It," pp. 138–65. Behrens's views on the relationship between technology and art can be found in a lecture that he delivered at the eighteenth annual meeting of the Verband deutscher Elektrotechniker on May 26, 1910, and is reprinted in *Peter Behrens and the AEG,* pp. 212–13.

GROPIUS, MEYER, AND THE BAUHAUS

A handsome, copious source for the Bauhaus is *The Bauhaus: Weimar, Dessauer, Berlin, Chicago,* ed. Hans Wingler (Cambridge: MIT Press, 1969). For more on Gropius and his views about modern technology-based culture, see my *American Genesis,* pp. 309–19. Gropius on mass-produced housing can be found in Gropius, "Wohnhaus-Industrie," *Berlin Tageblatt,* September 24, 1924; and in Herbert Gilbert, *The Dream of the Factory-Made House: Walter Gropius and Konrad Wachsmann* (Cambridge: MIT Press, 1984).

Gropius's successor at the Bauhaus presents his vision of the new modern culture in "Die neue Welt," *Das Werk* (1926): 205–24, reprinted in Bauhaus-Archiv, Berlin, *Hannes Meyer, 1889–1954: Architekt Urbanist Lehrer* (Berlin: Wilhelm Ernst & Sohn, 1989), pp. 70–73.

SULLIVAN AND WRIGHT

Louis H. Sullivan's dependence on Darwinism and his determination to design buildings related organically to their context is found in his *Autobiography of an Idea* (New York: Dover, 1924). Also see Carl W. Condit, "Sullivan's Skyscrapers as the Expression of Nineteenth-Century Technology," *Technology and Culture* I (1959): 78–93.

In 1901 Frank Lloyd Wright delivered a lecture at the Hull House in Chicago that came tantalizingly close to arguing that technological values should shape architecture: "The Art and Craft of the Machine," reprinted in *Roots of Contemporary American Architecture,* ed. Lewis Mumford (New York: Dover, 1972), pp. 169–85.

NEUE SACHLICHKEIT

A style of German painting flourishing after World War I mirrored the putative objectivity of science and technology. Helmut Lethen explains how the style embraced not only painting, but other realms of culture in *Neue Sachlichkeit, 1924–1932: Studien zur Literatur des Weissen Socialismus* (Stuttgart: J. B. Metzlersche, 1970).

Ingeborg Güssow focuses upon painting in "Malerei der Neuen Sachlichkeit," in *Kunst und Technik in den 20er Jahren: Neue Sachlichkeit und Gegenständlicher Konstructivismus,* ed. Helmut Friedel and Ingeborg Güssow (München: Städtische Galerie im Lenbachhaus, 1980), pp. 46–73.

DUCHAMP

Linda Dalrymple Henderson explores in depth Duchamp's numerous notes in order to analyze *The Large Glass* in *Duchamp in Context: Science and Technology in the Large Glass and Related Works* (Princeton: Princeton University Press, 1998). The influence of Duchamp among American artists is described in Francis M. Naumann, *New York DADA 1915–23* (New York: Harry N. Abrams, 1994).

SHEELER

Charles Sheeler's paintings, drawings, and photographs are shown and discussed in companion volumes: Carol Troyen and Erica E. Hirshler, *Charles Sheeler: Paintings and Drawings* (Boston: Museum of Fine Arts, 1987); and Theodore E. Stebbins Jr. and Norman Keyes Jr., *Charles Sheeler: The Photographs* (Boston: Museum of Fine Arts, 1987).

BOURKE-WHITE

The remarkable career and early industrial photographs of Margaret Bourke-White are interestingly presented in Stephen Bennet Phillips, *Margaret Bourke-White: The Photography of Design, 1927–1936* (New York: Rizzoli, 2003). See also Vicki Goldberg, *Bourke-White* (n.p.: United Technologies Corporation, 1988).

GIEDION

An encompassing overview of the history of design, especially modern American, is interpreted by Siegfried Giedion in *Mechanization Takes Command* (New York: Oxford, 1948).

LOEWY

Glenn Porter has ably described and illustrated the work of America's premier twentieth-century designer in *Raymond Loewy: Designs for a Consumer Culture* (Wilmington, Del.: Hagley Museum and Library, 2002).

ABSTRACT EXPRESSIONISM

Daniel Belgrad argues that an aesthetic of spontaneity spread among a host of artists, writers, and musicians, including bebop, immediately after World War II. Though not an organized cultural movement, it possessed, nevertheless, a loose coherence because of the interactions of its adherents in such places as Black Mountain, North Carolina; North Beach and San Francisco, California; and Greenwich Village, New York. Belgrad finds spontaneity provoked by corporate liberalism rather than by technological systems, as I argue. Daniel Belgrad, *Spontaneity: Improvisation and the Arts in Postwar America* (Chicago: University of Chicago Press, 1998).

Among the essays that I have drawn upon in arguing that abstract expressionism was, in part, a reaction against the oppressive controls of large systems are David Craven, "Abstract Expressionism, Automatism and the Age of Automation," *Art History* (March 13, 1990): 72–103; Clement Greenberg, "Modernist Painting," in *The New Art*, ed. Gregory Battcock (New York: Dutton, 1966), pp. 100–10; Meyer Schapiro, "The Liberating Quality of Avant-Garde Art," *Art News* 56 (Summer 1957): 36–42; and Meyer Schapiro, "Recent Abstract Painting," in *Modern Art, 19th and 20th Centuries: Selected Papers* (New York: G. Braziller, 1978), pp. 213–36. I am indebted to Amy Slayton for research from which I have drawn for this section and to Robert Brain for calling my attention to several relevant sources. See Brain's "The Algorithm of Pleasure," a paper presented at the Rathenau Summer Academy, Berlin, 1995.

CAGE

In presenting Cage, I have relied upon N. Katherine Hayles, "Chance Operations: Cagean Paradox and Contemporary Science," in *John Cage: Composed in America,* ed. Marjorie Perloff and Charles Junkerman (Chicago: University of Chicago Press, 1994), pp. 226–41. *Rolywholyover: A Circus* (New York: Rizzoli International, n.d.) is a boxed collection of essays about Cage as well as Cage's writings and letters. Reproductions of his works are included, too.

ARCHITECTS REACT AGAINST ORDER AND CONTROL

Denise Scott Brown and Robert Venturi, along with their colleague Steven Izenour, challenged the prevailing and dominating International Style of architecture that they found stifling the introduction of a less controlling and ordering architecture. Her argument can be found in "On Pop Art, Permissiveness, and Planning," *Journal of the American Institute of Planning* 35 (May 1969): 184–86; "Architectural Taste in a Pluralistic Society," *Harvard Architecture Review* 1 (Spring 1980): 41–51; and "On Architectural Formalism and Social Concern: A Discourse for Social Planners and Radical Chic Architects," *Oppositions* 5 (Summer 1976): 99–111. Robert Venturi, Denise Scott Brown, and Steven Izenour used Las Vegas architecture as an example of a vernacular form that could be transformed into a new architectural style in *Learning from Las Vegas: The Forgotten Symbolism of Architectural Form* (Cambridge: MIT Press, 1997).

ART, ARCHITECTURE, AND THE INFORMATION REVOLUTION

Herbert Muschamp, *New York Times* architectural critic, provides an example of computer-software-shaped design in "A Queens Factory Is Born Again, as Church," *New York Times* (September 5, 1999), Arts Section, p. 30. On Frank Gehry's use of computer software, see *Frank Gehry, Architect,* ed. Fiona Ragheb (New York: Guggenheim Museum, 2001).

Ecotechnological Environment

Anne Spirn called my attention in *The Granite Garden: Urban Nature and Human Design* (New York: Basic Books, 1984) to the overlapping natural and human-built systems found in cities.

MURCUTT

Glenn Murcutt's ecotechnological designs are described in E. M. Farrelly, *Three Houses: Glenn Murcutt* (London: Phaidon Press, 1993); and Françoise Fromonot, *Glenn Murcutt: Works and Projects* (London: Thames and Hudson, 1995). See also Glenn Murcutt, "Lecture," in *Technology Place & Architecture,* ed. Kenneth Frampton (New York: Rizzoli, 1998), pp. 56–70.

ECOTECHNOLOGICAL SYSTEMS

Richard White introduced me to organic machines in *The Organic Machine* (New York: Hill and Wang, 1995). His refreshing views on ecology can be found in "Are You an Environmentalist or Do You Work for a Living?" in *Uncommon Ground: Rethinking the Human Place in Nature,* ed. William Cronon (New York: W. W. Norton, 1995), pp. 171–85. Cronon writes of first nature and second nature in *Nature's Metropolis: Chicago and the Great West* (New York: W. W. Norton, 1991). See also his introduction to *Uncommon Ground,* pp. 23–56. A symposium at the National Academy of Engineering published its presentations in *Engineering and Environmental Challenges: A Technical Symposium on Earth Systems Engineering* (Washington, D.C.: National Academy of Engineering, 2003). The theme was the development of a new field, earth systems engineering, that deals with the natural and human-built worlds.

KISSIMMEE AND THE EVERGLADES

I have relied upon an unpublished essay on the Kissimmee project by Agatha H. Hughes, University of Pennsylvania. Michael Grunwald authored a series of informative articles on Kissimmee and the Everglades projects in the *Washington Post,* June 23–26, 2002. Helpful as well is Norman Boucher, "Back to the Everglades," *Technology Review* 98 (August 18, 1995): 24ff. I have also used the U.S. Army Corps of Engineers website.

James C. Scott alerted me to the negative side of large technological projects in *Seeing Like a State: How Certain Schemes to Improve the Human Condition Have Failed* (New Haven: Yale University Press, 1998), as did Paul Josephson in *Industrialized Nature* (Washington, D.C.: Shearwater Books, 2002).

PUBLIC PARTICIPATION

For an example of public participation in a large technological project, see my *Rescuing Prometheus* chapter on the Boston Central Artery/Tunnel project, pp. 197–254.

TECHNOLOGICAL LITERACY

Bill Joy raises doubts that the public can be adequately technologically literate in "Why the Future Doesn't Need Us," *Wired* (April 2000): 238–62. A counterargument can be found in Committee on Technological Literacy: National Academy of Engineering, *Technically Speaking: Why All Americans Need to Know More about Technology* (Washington, D.C.: National Academy Press, 2002).

Index